Dancing with the Vodka Terrorists

Misadventures in the 'Stans

Dancing with the Vodka Terrorists

Misadventures in the 'Stans

Rob Ferguson

**IGUANA
GEMS**

Published by Iguana Books
460 Richmond St. West, Suite 401
Toronto, Ontario, Canada
M5V 1Y1

Editor (of updated sections): Lisa Sparks
Book layout and design: Lisa Sparks, Greg Ioannou, Stephanie Martin
Cover design: Jane Awde Goodwin

Library and Archives Canada Cataloguing in Publication

Ferguson, Robert W.
    Dancing with the Vodka Terrorists : Misadventures in the 'Stans / Rob Ferguson.

Published in 2003 under title: *The devil and the disappearing sea: a true story about the Aral Sea catastrophe* ; published in 2004 under title: *The devil and the disappearing sea*, or, *How I tried to stop the world's worst ecological catastrophe*.
Includes bibliographical references.
Issued also in electronic formats.
ISBN 978-1-927403-24-2

    1. Ferguson, Robert W.--Travel--Uzbekistan. 2. Environmental degradation--Aral Sea Region (Uzbekistan and Kazakhstan). 3. Nature--Effect of human beings on--Aral Sea Region (Uzbekistan and Kazakhstan). 4. Uzbekistan--Description and travel. 5. Aral Sea Watershed (Uzbekistan and Kazakhstan)--Environmental conditions. 6. Asia, Central--Social conditions--1991-. 7. Murder--Uzbekistan. I. Ferguson, Robert W. Devil and the disappearing sea. II. Title.

DK948.867.F47A3 2012   958.7   C2012-907276-1

A version of this book was published as *The Devil and the Disappearing Sea* in 2002 (hardcover) and 2003 (paperback).

This is the original print edition of *Dancing with the Vodka Terrorist: Misadventures in the 'Stans*.

PRAISE FOR THE FIRST EDITION OF THIS BOOK

"...the cast of characters drives this narrative, a cast so odd, so uncooperative toward one another, so immune to reason and so venal, that they couldn't have been invented by Carl Hiassen or Elmore Leonard.... Luckily for Ferguson the writer, he finds himself another character in a very funny if depressing corruption-racketeering-kickback plot that ends in murder. To his credit, Ferguson tries to draw this shady group, his temporary colleagues, with some empathy, and never crosses into outright parody. Ferguson has crafted something unique, something that blends memoir and documentary with noirish satire, and manages its shifts in tone very smoothly."
– *Globe & Mail*

"...eminently readable, held together by the author's careful research and attention to detail. As Ferguson travels through the region, he details its bold, colourful history of intense beauty and violent conquest, of Silk Road havens and Soviet brutalism. He vividly describes cultures in perpetual flux and ancient ideas concerning everything from religion to irrigation. He also has a keen eye for character. Team members are evoked memorably; toadying yes-men, corrupt apparatchiks, even the occasional idealist and loyal friend. Key figures, like Mr. G. and Shakhlo Abdullayeva, the team's office manager, are particularly well-drawn."
– *National Post*

"The story of the Aral Sea, the most stupefying and outrageous ecological disaster in human history, is the central narrative ... [of] Rob Ferguson's [book] ... it's an edge-of-your- seat ride [and] ... has the pacing of a detective novel and characters out of Evelyn Waugh. Who would have thought a first-time author could pull off something like that?"
– *Georgia Straight*

"If Hollywood ever makes a movie based on The Devil and the Disappearing Sea, Rob Ferguson would like Kevin Spacey to play him. Such a film would be a taut thriller."
– *Victoria Times-Colonist*

"The pith of Ferguson's fascinating debut–a hybrid of sightseeing travelogue, political history lesson, dire ecological warning and unsolved murder mystery–is that the Aral Sea, once the fourth-largest inland body of water on Earth, is shrinking fast.... His wry account of a turbulent year clearly articulates the tragic consequences of what he now deems inevitable failure ... Readers will finish the book knowing with certainty why the Aral Sea disaster has been described as a slow-motion Chernobyl."
– *Publishers Weekly*

"... this absorbing real-life thriller ... reads like novel, though if it were fiction, it would be billed as a black comedy ... The story is so absurd it would be funny if it were not about such an important issue. In this, his first book, Ferguson merges statistical information into the flow of the book with ease. He also details the history of the region, providing useful background to the events of his 11 months there. He describes each of the places he visits with a traveller's practised eye, providing an engaging travelogue on the legendary cities of the Silk Road ... somehow he managed to maintain his dedication and sense of humour, resulting in this informative and compelling book."
– *Winnipeg Free Press*

Best book of the year, *Outpost* magazine:

"A weird, entertaining and informative amalgam of a book: part memoir of Ferguson's time as an NGO worker in Uzbekistan, part ecological nightmare, part satire, part absurdist noir. From the unavoidable tragedy of the disappearing Aral Sea, to the machinations of a corrupt bureaucracy, Ferguson finds himself in one bizarre scenario after another, always with his wit intact."
– *Outpost* magazine

For my parents,
Margaret and Bob Ferguson of Winnipeg

But the majestic river floated on,
Out of the mist and hum of that low land,
Into the frosty starlight, and there moved,
Rejoicing, through the hush'd Chorasmian waste,
Under the solitary moon; – he flow'd
Right for the polar star, past Orgunjè,
Brimming, and bright, and large; then sands begin
To hem his watery march, and dam his streams,
And split his currents; that for many a league
The shorn and parcell'd Oxus strains along
Through beds of sand and matted rushy isles –
Oxus, forgetting the bright speed he had
In his high mountain-cradle in Pamere,
A foil'd circuitous wanderer – till at last
The long'd-for dash of waves is heard, and wide
His luminous home of waters opens, bright
And tranquil, from whose floor the new-bathed stars
Emerge, and shine upon the Aral Sea.

– from "Sohrab and Rustum: An Episode"
by Matthew Arnold

And He it is Who sends the Winds as heralds of glad tidings,
going before His Mercy, and We send down pure water from the sky,
That with it We may give life to a dead land, and
slake the thirst of things We have created, – cattle and men in
great numbers.
And we have distributed the (water) amongst them, in order that they
may celebrate (Our) praises, but most men are averse (to anything)
but (rank) ingratitude.

– The Qur'an, Surah 25:48–50

AUTHOR'S NOTE

This book, originally published as *The Devil and the Disappearing Sea*, is based on actual events that took place in the five former Soviet republics of Central Asia during the year 2000. All persons identified are real and carry their own names, except for a few minor characters whose names I was unable to recall. This is a nonfiction work, and the events and narrative have been recreated as faithfully as possible, through recollections, journal entries, research and discussions with my former co-workers, to whom I am greatly indebted. The dialogue aims to capture the essence and significance of the conversations at the time, but as it was not recorded and was sometimes filtered through an interpreter, it may not comply exactly with the speakers' words. I also took a few liberties regarding the timing of events. However, my intention has been to remain as true as possible to the characters and the quintessential flow of events as I experienced them. In 2012 the original version was revised slightly, an introduction and epilogue were added, and it was given a new title.

Contents

Introduction

Eleven years ago I came back from a year working on a development project in the 'Stans of Central Asia thrilled by the adventures, or misadventures, I'd just escaped from. As I related what had happened to me in the five former Soviet republics to friends and family, I realized it was not only a great story but would make a terrific book. It had all the necessary ingredients: an exotic setting including the old cities of the storied Silk Road, that ancient crossroads between Europe and the Far East, the soaring Heavenly Mountains and the fertile valleys that once were the fruit basket of the Soviet Union, larger-than-life characters "that couldn't have been invented by Carl Hiassen or Elmore Leonard," as The Globe and Mail reviewer wrote, and a plot that climaxed with a murder in which I was a prime suspect. There was even a car chase.

In 2003 Raincoast Books published it as *The Devil and the Disappearing Sea* and promoted it mainly as a book about "the world's worst ecological catastrophe." As a result, in bookstores it typically got slotted on a lower shelf with a small collection of earnest environment books where, alas, fans of travel adventure, real-life thrillers and black humour were sure not to find it. It got very limited distribution in the United States, none at all in the UK and Continental Europe where the Aral Sea registers more than just the answer to the crossword clue, "shrinking sea in Central Asia." There, due to Russia's continual political games over the sale of their natural gas to Europe, a greater consciousness of environmental disasters and the region's closer proximity, the 'Stans exist as real countries on the map. Unsurprisingly my book had no official distribution in Russia and Central Asia, although I got reports copies were circulating around the 'Stan's expat community who found it reflective of their own experiences, and that a rough Russian translation was also covertly making the rounds.

In recent years Sacha Baron Cohen's mockumentary *Borat: Cultural Learnings of America for Make Benefit Glorious Nation of Kazakhstan*, and Gary Steyngart's *Absurdistan*, both his book and the movie, have helped to put the 'Stans if not exactly on the map then into wider consciousness, while at the same time stereotyping them as bizarre and wacky states stuck in a time warp. From my experience this is not altogether inaccurate. Although often, when I talk about having worked there, many people still draw a blank, and when I mention Kazakhstan, which is the best known 'Stan in the region, a few have even showed surprise and commented they thought it was "a made-up country."

"If only," I respond.

Re-reading my book I found myself once again immersed in those frustrating, infuriating and never-a-dull-moment games of my former bosses, those gregarious and scheming apparatchiks who delighted in making westerners squirm. That part of the world has hardly changed in the years since I left: the Aral Sea, although not completely disappeared, has shrunk even more, the region's autocratic presidents – two are still in power – continue to repress their citizens, the state-controlled media still disseminates outrageous propaganda that pretends their five nations are at the forefront of economic progress and modernity, and no doubt the World Bank, international donors and aid organizations continue to pay off the bosses in hopes of achieving perchance a little progressive development.

I found myself laughing again at the corny shtick of my immediate boss Bozov and his sidekick Nadir – you can't make up more terrifically evocative names than that – wily big boss Mr. G who looked and acted so much like a monster he even growled, devious Shakhlo with her transparent charms, "specialists" Tulenbai and Abbazbek and their Tweedledum and Tweedledee routine, matronly "pretty-as-a-spring-flower" Valentina and her Escher-like flow charts, sentimental drunk Bayalimov who now and then would speak the blunt truth, diminutive Anatoly of the World Bank with his chirpy lies and bad clichés and those two impossibly jolly consultants named Frank. Looking back, it seemed like a madcap comedy farce that had all been choreographed: the tongue-in-cheek jokes, the sneaky tricks, the propaganda campaigns, the sabotaged workshops and Bozov's recurring slyly ironic declaration: "Robert, the people of Central Asia are very complicated!"

Bozov was right about that, although it wasn't so much the people as just things that were so complicated. These complications were born of a culture nearly completely lacking in trust. Most people operated out of fear and self-interest, jumping at any opportunity to skim a few dollars from the absurdly lucrative funds of international aid projects so that they could pay a bribe to the principal of a decent school to get their child enrolled, or repair the old Lada or just put food on the table. It was, and I'm sure still is, a culture of desperation and paranoia. It was cynical, manipulative, sycophant-inducing and lacking in any sense of basic morality. I recall Jean-Charles Torrion, the director of the French consultancy that won the contract for our public awareness component of the Aral Sea project, saying in irritation, "Never was there any cooperation.... Only games. I should not have signed this contract.... Never was there good faith on their side!" It brought out the worst in nearly everyone, which was why anyone who stood up to the corruption and refused to play the games was remarkably brave.

Victor and Sabit were my salvation. They are the heroes of this tragedy, albeit a black comic tragedy, not only because we failed in our overly ambitious attempts to create effective public awareness campaigns that persuaded water users' to save water, but because our project only exacerbated this culture of corruption and mistrust and ultimately caused a gruesome murder. Victor Tsoy was steadfast in his refusal to be cowed by the sleaze. I don't know how many times I sat waiting in his car at checkpoints while the police harassed him, expecting a small bribe he would never pay. He stood up to Mr. G and got fired for it and several years after our project ended may well have sacrificed his life for his principles (see the Epilogue). Sabit Madaliev was just as ethical – "Robert, they are all such terrible liars!" – but he also knew how to cleverly play off the bosses to our advantage, and whenever I grew exasperated and ready to stoop to their sly demands, he would click his tongue at me and remind me of my moral duty. They were both exceptional, and I am deeply indebted to them.

It was Shakhlo's grisly homicide that shocked me back into reality after nearly a year of playing maddening, and at times, very entertaining games. What happened to her sickens and shocks me yet today, and the fact that no one has ever been convicted of her murder again shows how unjust and lacking in integrity things really are in that part of the world. I recall somebody in Tashkent telling me it cost $1000 to have someone knocked off, a price that may well have been high. Factoring in police incompetence, or paying them off if necessary, it was certainly an easy and cheap solution when someone has offended you. What strikes me now is how expendable Shakhlo was. Whether Bozov and Mr. G were complicit or not, it's very clear she was merely a pawn in their schemes. In the end she didn't really matter to them at all.

I'm left with the feeling that no one in this tragic-comedy was completely innocent. It's clear now we were all abetting in a crime, which was working on a project that despite its excellent intentions – almost all international projects have good intentions – was executed as a farce. The games, the tricks, the lies, the payoffs, the blackmail and even the murder were all part of doing business in the 'Stans. Although this is not to say I would not go back there again, or to Africa or Asia or South America, to work on similarly "complicated" projects. Because I believe that eventually, somehow, things do change for the better.

Toronto, August 15, 2012

Prologue

On December 8, 1999, a cool fall day in Tashkent, Uzbekistan, a boyish-looking Frenchman and a giant bald Uzbek of Tartar descent signed an agreement that initiated a one-year training mission. Jean-Charles Torrion, the director of development for BDPA, a Paris-based international consultancy group, and Rim Guiniyatullin, the head of a project aiming to save the Aral Sea, couldn't have been more different. Torrion, in his early forties, was an easygoing, urbane consultant with experience working in dozens of countries, from Argentina to Mali. Guiniyatullin, in his sixties, was a former Soviet apparatchik, a shrewd irrigation expert with a fierce temper and no love for the West.

Torrion had just agreed to send a group of communications specialists to Central Asia, where they would train five national teams to raise awareness about one of the world's biggest environmental catastrophes: the rapid disappearance of the Aral Sea. According to the contract, his specialists were to develop "communications capacity, undertake opinion research and analysis, build a long-term communications strategy, and advise upon and help implement critical public awareness activities." Torrion hired me as team leader. I would coordinate our activities and manage our team and resources. I was thrilled to be part of it, and to be heading for an exotic corner of the world that remained a dark mystery to most Westerners.

The new US$250 million project I was to work on, the Water and Environmental Management Project for the Aral Sea Basin (Aral Sea project for short), was funded by the Global Environment Facility and monitored by the World Bank. Its goals were to upgrade the massive crumbling irrigation system in Central Asia that nurtured millions of hectares of cotton fields and starved the Aral Sea of water. The huge inland sea is rapidly disappearing, leaving behind a toxic desert that is poisoning any remaining life forms. Oddly, this disaster was being largely ignored by the West – and by most Central Asians. Our mission was to bring the disaster to the people, and ultimately to save water. If consumption continued at its current rate, crippling water shortages would soon be widespread.

"Public awareness" refers to campaigns that are run by governments or nonprofit groups in an effort to broaden public perceptions of major community issues, such as the dangers of AIDS or smoking. The goal is to stimulate positive changes in behaviour. Like advertising, public awareness campaigns target affected groups of the population with call-to-action messages or slogans: "Practise safe sex!" or "Second-hand smoke kills." I had just spent two years in Mongolia with the United Nations Development Program training environmental groups to carry out public awareness activities that aimed to protect endangered species, recycle waste and prevent brush fires. The response had been enthusiastic and the newly democratic government supportive. But the five former Soviet states of Central Asia were dictatorships and after decades of Soviet propaganda and strict controls over the media, the people were repressed and skeptical. They were unfamiliar with Western-style public participation, cynical about public-spirited messages and felt little responsibility for their governments' actions. My new mission was going to be a challenge.

The agreement Torrion signed that day in December was worth US $670,000, not a large amount by the standards of international development contracts. But it was still a significant feat, the first contract awarded to Western consultants under the Aral Sea project. As the World Bank would later admit, we were the guinea pigs.

1   Under Suspicion

Compared to the gulag of rooms around him, Numan Karimov's office was a triumph of glasnost. New textured wallpaper masked the sallow green that pervaded the rest of the Chilanzar microdistrict police station. Three large posters of turquoise and gold domes set against azure skies advertised SAMARKAND, BUKHARA and KHIVA. The furniture was new: vinyl wood grain with chrome trim. The floor was dappled linoleum, slightly spongy underfoot. The overhead light was a glaring fluorescent.

"It looks like a travel agent's office," I told him with a polite smile. I was sitting with Ira Vovchenko, my interpreter, at an interrogation table pushed hard against Karimov's desk. The deputy chief of homicide for the Tashkent police laughed at Ira's translation of my joke as he pulled a dangerous-looking electric prong out of a pot of boiling water.

"I did it myself," he announced, pleased with himself.

I thought he seemed a little too cheerful for a homicide investigator. In his early forties, Karimov had an oily complexion, thick black hair and sharp eyes. He had shunned the suit-and-tie uniform of his fellow detectives for jeans and a green polo shirt open at the neck.

"The police have no money for interior decoration," he said with a little smirk.

He set two cups of tepid water, a jar of Nescafé, a tarnished spoon and a box of crusty sugar cubes on the table in front of us. Ira and I mixed our coffees. Karimov sat down at his desk and eyed his mobile phone, the only item on his desk. My mobile phone was costing me about US$40 a month, about as much as his "official" salary, I guessed.

"It's for my parents," he said with a smile when he saw me looking at it. "They're old and sometimes they need to call me." Then his expression turned serious. "I must ask you some questions about Shakhlo Abdullayeva."

He opened a drawer and searched for something. I glanced at Ira. She sat poised and rigid, leaning with her elbows on the table. Her auburn hair was newly coiffed in a perky bob that accentuated her severe appearance. She was very pale.

"I understand you worked with Mrs. Abdullayeva."

"She used to be my employee."

He pulled out a file folder and put it on his desk.

"What was your relationship?"

"She was my office manager. She did the books, bought supplies, arranged visas and housing for our consultants."

He studied me again. I was expecting the question, "Where were you the morning of Saturday, December 16?", but instead he asked, "How would you describe her?" I thought it was an odd question.

"About forty, black hair, quite attractive." I wanted to add that she had a bewitching smile, full of ironic fun, that she danced like a belly dancer, flirted outrageously and could laugh recklessly. "But you would know what she looked like," I said with slight provocation.

"The body was badly mutilated," he replied with a little smile.

"What about her personality?"

"She could be charming, but at other times difficult. Sometimes she wasn't completely honest." I smiled to imply this would be no surprise to him.

"Really?" said Karimov once Ira had interpreted this statement.

He acted as if he'd not heard this before. But he looked pleased. "Did she steal from you?"

"From our training team."

"How much?"

"She skimmed money off some of our business transactions."

"Did you report this to the police?"

"No. We settled the matter internally."

He eyed me doubtfully, then opened the file folder and looked at its contents. He wasn't writing anything down and I thought it a strange way to conduct an investigation.

"What is your work here in Tashkent? Your project?"

His voice was taunting and he was looking at me with a mixture of admiration and disdain. We were about the same age and I guessed he approved of my easygoing nature and envied my well-paid job, my Canadian passport and my freedom. But he disdained the expat lifestyle: the expensive imported food from the Ardus Supermarket, the draft Holsten Pilsners with other expats every Friday night at the Salty Dog. He could not guess that I found the expat scene tedious and was not a regular at the Salty Dog.

I told him I was team leader of a training mission that was part of the public awareness component of the Aral Sea project, which was overseen by the World Bank. He narrowed his eyes, considering this. "I coordinate the missions of international consultants who come to Tashkent and tour the region. We provide support for local public awareness officials who give out information on the Aral Sea catastrophe." He continued to stare at me, unimpressed and perhaps a little bored. "The goal of the public awareness component is to persuade the people of Central Asia that water has to be saved so that the Aral Sea can be saved. As you know, the Aral Sea is rapidly disappearing due to the over-irrigation of cotton crops."

He laughed loudly when he heard this. I wasn't sure if he thought it was funny that the Aral Sea should be saved or that a foreigner should be trying to explain the Aral Sea catastrophe to him. Or both. He raised his bushy eyebrows facetiously.

"Is your work successful?"

I wanted to say no, not at all, that we were failing because the project bosses were refusing to cooperate and only playing games with us, and that the World Bank officials who were overseeing the project were being less than helpful. Instead I said:

"It takes years to change attitudes. But we're getting the process on the right track."

Ira didn't flinch as she translated this.

"I'm glad," he said cheerfully but insincerely. He continued talking but Ira abruptly stopped interpreting. She looked upset. After a pronounced pause she relayed a shortened version of his words:

"I understand that you and Shakhlo Abdullayeva were lovers."

I said nothing for a few seconds. Then I smiled back sarcastically.

"Who told you that?" Karimov didn't answer. "I'm sorry to disappoint you, but we had a strictly professional relationship."

He sneered at me for a few long seconds.

"Isn't it true that you accused her of stealing money from your safe and then fired her when you found out she was having an affair with one of your consultants?"

In the car Ira was eerily quiet, locking me out of whatever thoughts Detective Karimov had provoked in her. When he was done with me, he had questioned her alone for half an hour. Maybe he'd managed to breach her trust in me. That would be his technique, I decided, my paranoia growing: divide and weaken until a confession rolled out. He needed a murderer and one way or another he would get himself one.

9

Ira asked our driver to drop her at her flat. As she got out I suggested she take the next day off. She said that she would call in the morning.

When I got home, I called Lufthansa and asked the agent to change my booking to the first available flight out. He managed to get me a seat on December 21. Karimov had told me not to go anywhere, but in forty-eight hours I intended to be on a flight to London.

2   But There Is No Lemon!

Eleven months earlier I was gazing out the porthole window of Turkish Airlines flight 1862. Below me was a collection of blue-green splotches rimmed in white that caught the light and shimmered amid the greys and yellows of the Kyzylkum Desert. The sun was setting, casting long shadows across the scorched landscape, empty of signs of humanity except for a few vehicle tracks skirting the wasteland like the trails of crabs on a beach. According to the map of our flight path in the on-board magazine, I figured we were somewhere over the delta of the Oxus, a river that once emptied into the Aral Sea and that was eulogized so memorably in Matthew Arnold's epic poem "Sohrab and Rustum: An Episode": "The shorn and parcell'd Oxus strains along/Through beds of sand and matted rushy isles ... " But these days, the Aral had shrunk into a hodgepodge of shallow lakes so salty they poisoned rather than sustained life and the Oxus, now known as the Amu Darya, barely contributed a drop.

I was flying to Tashkent, the capital of Uzbekistan, to begin my one-year mission as leader of a team of Western consultants. We were all experienced in applying methods that aimed to change public opinion on serious community issues. In this case it was the dire need for water conservation. We would work under the public awareness component of the Water and Environmental Management Project for the Aral Sea Basin – the Aral Sea project – a regional project overseen by the World Bank.

I had been waiting more than a year for the job to start; it had seemed jinxed all the way. First the bidding process had been delayed by the terrorist attacks in Tashkent in February 1999, blamed on Muslim extremists. Then later that year, after BDPA had won the contract, the head of the Aral Sea project had held things up further by bickering over the terms. When I finally got the word to come, I was ecstatic. At last I was to work on a project that was striving to contain a huge environmental crisis, *and* explore an exotic and mysterious part of the world.

The desolate panorama below had a three thousand year storybook history: Alexander the Great and Chinggis Khan (Chinggis is the Mongol spelling of Genghis) had conquered it and khans and Cossacks had slaughtered, looted and traded on its Silk Road. Now it was making history again as the scene of one of the 20th century's biggest environmental disasters, a man-made tragedy of epic proportions that had only been exposed in the late 1980s as Gorbachev's reforms loosened controls over the Soviet media.

In 1960 the Aral Sea was 66,900 square kilometres of crystalline, lightly salted water. But by the end of the century its level had dropped 19 metres, it had lost 80 percent of its volume and more than half its surface area, and its salinity had increased to 40 percent. As it retreats it is leaving an immense plain of toxic devastation. The disappearing sea, once the world's fourth-largest inland body of water, has triggered a full-blown environmental disaster by ravaging ecosystems, creating a poisonous dust that was killing thousands of inhabitants prematurely, and producing extreme weather, crop failures and polluted water supplies. And despite all the warnings, charged rhetoric and foreign aid, the catastrophe crept unabated into the 21st century. By 2000 it was affecting the whole Aral Sea basin, 1.5 million square kilometres of snow-capped mountains, fertile alluvial valleys and vast steppe and desert that touched six nations: Kazakhstan, Kyrgyzstan, Uzbekistan, Tajikistan, Turkmenistan and Afghanistan. The Aral Sea disaster was denigrating living conditions for more than forty million people.

(By way of comparison, picture Lake Huron, slightly smaller at 59,600 square kilometres, shrinking until Georgian Bay is a boggy wasteland dotted with stands of coniferous forest and abandoned vacation homes, and what remains of the lake is a wide channel flanked by vast mudflats.)

As I watched the disaster zone fade into night, I resolved to visit this wasteland, to see it up close. I wanted to touch the brackish waters of the shrinking Aral, walk along the sea's vast shores and smell its putrid air. I wanted to talk to the suffering inhabitants and hear their stories. I wanted to take pictures of the rusting fishing boats beached permanently in the dunes, the devastated settlements, the farmers standing amid their failing crops of cotton and rice. This is what had brought me here and I was determined to experience it.

"Robert!" shouted Kadirbek Bozov, slamming down the phone. "You are finally here!" He was grinning at me and speaking in Russian. Victor Tsoy, my new right-hand man, was translating. "You were supposed to be here January 9!" (It was now January 24.) "Where were you? What were you doing? I was waiting for you every day!"

Bozov, the director of the public awareness component of the Aral Sea project, was half-chastising, half-teasing. Short and pudgy, he was dressed in a wrinkled suit, white shirt and striped tie. He looked like a doughy middle-aged schoolboy. We were in his office, down a murky corridor on the seventh floor of the Brezhnevian customs police building, where the project leased two floors. The customs police were a branch of the Uzbek KGB.

Nadir Maksumov, Bozov's assistant, jumped up and shook both our hands.

"You're welcome!" he shouted. "Very nice to meet you! You're welcome!" Then he sat down and laughed as if speaking English were a great joke.

"*Chai! Chai!*" yelled Bozov, and Nadir jumped up again to make tea.

Their two desks were pushed together so that they faced each other. A Chinese-red phone floated between them on top of stacks of bulging file folders. On adjoining tables their monitors displayed open Word files in Russian. The walls were covered in posters and calendars lamenting the Aral Sea disaster. Above Bozov's desk was a map of Kyrgyzstan, his homeland.

Bozov motioned for me to sit on a padded chair close to him. He pointed Victor to a wooden chair against the wall. He took my right hand in both of his and massaged it.

"I couldn't get an Uzbek visa in Canada," I explained. "So I flew to Paris and managed to get one there. The Uzbeks don't make these things very easy." Victor translated this, but Bozov waved his words away as if none of this mattered now.

Victor was 43, a handsome man with a doleful frown and a big moustache lightly flecked with grey. He had just spent several months assisting Jean-Charles Torrion, my contract boss with the French firm BDPA, who had hired me, with the tricky contract negotiations. When they finally got their deal, JC had rewarded Victor by hiring him as my local counterpart. His job was to guide and support me in managing our training team. Victor was a media specialist and ran his own nongovernment organization (NGO), Rabat Malik, which specialized in ecological education for children.

"I was just talking with Valentina Kasymova," said Bozov, "the leader of our Kyrgyz team. She is the best public awareness specialist in Central Asia and beautiful as a spring flower in the mountains! She sends you her greetings! She says she will see you soon!"

13

In the 1980s Bozov had participated on regional councils that regulated water use through the complex system of dams, reservoirs and canals that the Soviets had built up during the post-war years. The end of the Soviet Union chopped the region into five new states, making the sharing of water complicated and highly politicized. Bozov became a frequent Kyrgyz rep on regional water matters. In 1997 the Kyrgyz government selected him to head the "Kyrgyz component" of the new Aral Sea project. The five states were each allocated responsibility for one component: Component A was Kazakh – water and salt management; Component B, Kyrgyz – public awareness; Component C, Tajik – dam safety and reservoir management; Component D, Turkmen – transboundary water monitoring; and Component E, Uzbek – wetland restoration. As if formulated by a logician, the project operated on base five: five states and five components. This seemingly rational framework, I later realized, provided a screen for some highly convoluted politicking.

By early 1999 Bozov had recruited some old cohorts to lead three of his national public awareness teams, each charged with disseminating information on the Aral Sea disaster and trumpeting the project in their home states. The new leaders of the Kazakh, Tajik and Kyrgyz (headed by Valentina, just on the phone) teams had all worked with Bozov before and shared a similar mindset. The independently minded Uzbeks held a competition to select their public awareness team and the Turkmens just put the whole thing off. Reluctant participants in most regional activities, the Turkmens played by their own rules, or didn't play at all.

The red phone rang and Nadir grabbed it, shouting, "*Aloa, aloa!*" He recognized the caller and exchanged cheerful greetings. Then he passed the phone to Bozov who, lighting a cigarette, bellowed more happy salutations.

One of the posters on the wall caught my eye. A young woman draped in long psychedelic dress and head scarf was propped on a bluff overlooking a vast mountain lake. She was superimposed, a chaste calendar girl plunked crudely into the scene. She looked like a Central Asian Mona Lisa and I thought her benign smile unintentionally ironic. I guessed the water behind her was supposed to be the Aral Sea.

The Aral Sea was born about 35,000 years ago, at the end of the last ice age, when the mammoth sheets of ice that had been suffocating the northern hemisphere finally melted. They receded, exposing a land of multi-coloured bogs and seas, which gradually drained into the depression in the middle of the planet's biggest landmass. Two rivers, rising from glaciers in the towering mountains to the south and east, flowed across steppe and desert into this huge inland sea. Grasses flourished on the steppe; reeds, shrubs, elms and poplars filled the riverbanks and swamps; and walnut, pistachio, apple, apricot and juniper groves multiplied in the valleys and desert oases. Summers in the foothills brought wild tulips, poppies and flamingos. Long-horned sheep, yaks and marmots roamed and scurried over the higher tablelands and the mountain valleys. Antelope, pheasants, falcons and hawks inhabited the lower plains while gazelle, jerboas, vipers and leopards stalked each other on the parched red deserts.

Around 5000 BCE nomadic cattle herders from the northwest began roaming the area, battling each other in chariots and grazing their horses and bullocks in the abundant grasses. About this time the first settlers began to till the fertile soils along the Amu Darya (or Oxus) and Syr Darya rivers, creating water furrows and drains that allowed them to cultivate melons, apricots, apples, grapes, wheat and cotton. Their descendants today are the Tajiks. Irrigation, if not actually invented here, was certainly practised skilfully. By 500 BCE the region had evolved into three kingdoms: Khorezm, at the delta of the Amu Darya next to the Aral Sea; Sogdiana, the lands between the Amu Darya and Syr Darya that the Romans called Transoxiana; and Saka, the steppe beyond the Syr Darya as well as the mountain areas to the south and east. These regimes thrived, were conquered and then rose again, in a cycle repeated over and over.

In 329 BCE Alexander the Great led his Macedonian armies into Marakanda, known today as Samarkand, then crossed the Syr Darya. After an 18-month siege he conquered Sogdiana and added the beautiful Bactrian princess Roxana to his collection of wives. He also opened the region to European trade and ushered in an enlightened age. Shortly thereafter warring tribes from the Far East also discovered the region. But after the inevitable plunders and slaughters they too settled down to business and the Silk Road was born. Agriculture flourished, enhanced by the interchange of technologies that arrived via the new East-West trade routes between China and Europe. Ideas, religions and cultures mingled, fused and gave birth to new art forms, scientific advances and technical innovations. It was a tolerant, striving society. Its more astute traders amassed fortunes and the opulent caravan towns of Kokand, Samarkand, Bukhara and Merv prospered.

15

But soon the spirited tribes of the steppe began to raid the three affluent kingdoms. Trade dwindled with the decline of the Chinese, Roman, Iranian and Kushan (in northern India) empires, and the sedentary farmers with their irrigated plots and the townspeople with their banking and bartering found themselves vulnerable. The Blue Turks charged in from southern Siberia. The Huns attacked from the north. Then Islamic armies from Arabia overran the region and conquered much of it by 700 CE. Things finally calmed down again by the ninth century when the Sunni Muslim Samanid dynasty ruled peacefully and allowed the Persian and Turkic cultures to flourish. *Bukhoro-i-sharif*, "Noble Bukhara," emerged as a major centre of Islamic culture and its eminent scholars and scientists turned it into Central Asia's Islamic heart.

But a number of unrelenting disputes devolved into a string of wars that ultimately killed this golden Islamic age. Then in 1220 the scourge of the eastern steppe, Chinggis Khan, charged in at the head of his Mongol Hordes. They terrorized, plundered and butchered. Bukhara was burned to the ground. Samarkand was sacked and all its citizens slaughtered. For those that survived, life under the Mongols eventually settled down, though the region would not fully recover for more than 600 years. Silk Road trade slowly revived and became legendary. On the Khan's death the region was carved up among his sons, whose empires slowly fractured, allowing the Turkic people in the region to rise. Tamerlane, who hailed from a minor Turkic tribe near Samarkand but claimed a lineage to Chinggis, followed the Khan's example and pillaged most of greater Central Asia. Tamerlane is a man of myth and mystery. Also known as Timur the Lame, he walked with a limp, possibly because of a war wound. With the plundered riches he rebuilt Samarkand into his showcase capital – much of the resplendent architecture evident there today dates from this era. (Since the collapse of the Soviet Empire, Chinggis and Tamerlane, both reviled as barbarians by the Soviets, have been reborn as national heroes in their respective homelands, Mongolia and Uzbekistan.)

"Sugar?" asked Nadir cheerfully as he poured a small amount of tea into a bowl for me. "*Limon*?" He made a sour face.

In his mid-thirties, he was wearing a roll-collared cardigan with a puttering-around-the-house look that he periodically tightened over his slightly protruding belly as though it had just filled out overnight and he wasn't yet used to it. His expressions ranged from mock grins exposing a set of gold teeth to ironic gapes full of feigned shock. He was a ham.

"*Limon*," I said, "*pazhalsta*" (please in Russian).

"Lemon?" Nadir looked at Victor and gagged. "But there *isn't* any lemon!"

I glanced at Victor, whose amused frown implied that Nadir was a harmless fool.

Bozov banged down the receiver and again took my right hand in both of his.

"Robert, I must advise you on some matters." He glanced at Victor for translation, picked up his cigarette, inhaled deeply and blew the smoke half into my face. Then he eyed me over large reading glasses that rested halfway down a wide nose that ballooned between chubby flushed cheeks. "You and I are both foreigners in this country and we must respect Uzbek laws and traditions. You see, the people of Central Asia are very complicated."

"*Very* complicated," added, Nadir nodding.

"That's why you must listen to me," continued Bozov. "You don't understand the people of Central Asia. But I will be your guide. I will show you how to be successful. Together we will be successful, but you must do exactly what I say."

"You must *always* do what Mr. Bozov says!" echoed Nadir.

By the 16th century the Uzbeks, as the sedentary farmers and traders were now called, ruled Transoxiana, the lands between the Amu Darya and the Syr Darya, principally today's Uzbekistan. The Kazakhs, the horse-riding herders of the steppe, ruled one of the world's last nomadic empires on the vast grassland north of the Syr Darya, more or less today's Kazakhstan. The differences between the region's two dominant ethnic groups are still evident today. The Uzbeks, intense tillers of the soil, were smugly administrative, urban and literate, and earnestly Islamic. The Kazakhs, and their mountain cousins the Kyrgyz, were free-spirited horsemen, tribal and pastoral and disinclined towards strict Islamic codes. Uzbek lands were ruled by three khanates: Kokand in the Fergana Valley, Bukhara in the central region, and Khiva in the ancient Khorezm territory at the delta of the Amu Darya on the Aral Sea. The Kazakh steppe was controlled by three Mongol-descended hordes: the Great Horde on the steppe adjacent to the Tian Shan (Heavenly) Mountains; the Middle Horde on lands to the east of the Aral Sea; and the Little Horde on the plains north of the sea.

17

Enter the Russians. By the mid-19th century Central Asia, with its depraved despots, feudal khanates and tribal nomadism, seemed a good bet for the tsars' new imperialistic designs. The region was abounding in profitable goods, especially cotton, silk and wool. Britain, France, Holland, Spain and Portugal were conquering foreign lands and creating global empires and Russia wanted in on the action. With Britain already in India, St. Petersburg's arrogant young military officers were keen to probe the exotic hinterlands of Russia's soft Asian underbelly and extend their southern frontier. The Kazakhs were first to be conquered. By 1848 a force of Cossacks had mostly eradicated the Three Great Hordes and were colonizing the Kazakh steppe and taiga. The Uzbeks were next. Bozov's Kyrgyz ancestors in the Tian Shan Mountains, feuding among themselves, played a minor role in the conquest of Russian Kokand in the fertile Fergana Valley. Tashkent fell in 1865, initiating a Russification that ended only in 1991. Bukhara soon succumbed, then Khiva in the Amu Darya delta, where Russians had been routinely sold into slavery. All became protectorates of the tsar. The unaligned tribes on the Turkmen steppe towards the Caspian Sea stubbornly resisted until 15,000 were slaughtered in a bloodbath in 1881. By the end of the century, the Russians had carved out a formidable empire in Central Asia.

Bozov leaped from his chair to hug a wild-eyed man with a pockmarked face standing in the doorway. Nadir also hugged him. Bozov introduced him as an old crony and member of the Kazakh public awareness team.

"The Kazakhs have the very best team," raved Bozov.

I shook the man's clammy hand and he wouldn't let go. His vodka breath wheezed in my face.

"Very important that you must to be Mr. Bozov friend!" he said in English.

I glanced at Nadir, who rolled his eyes.

He was the first of a series of men who filed in that morning; Bozov had apparently ordered them all to come. After handshakes and salutations they lined up on the wooden chairs next to Victor, eight in total. They were all "water specialists," all dressed in slightly mismatching suits, variations on brown pants and dark blue jackets, and all in their forties and fifties. (We were supposed to be discussing the hiring of my training team, which was to include local *media* specialists.) They smiled unctuously and carried a sluggish air of long-term unemployment. Bozov gushed about who was the best at flow utilizations and who knew effective drainage techniques while I grasped that Bozov equated public awareness with expertise on irrigation rather than mastery of communication skills.

I frowned at Victor, who frowned back.

The last decades of the 19th century were the era of the Great Game, a stealthy cold war between the British and the Russians played out among the shifting and murky frontiers of Central Asia. The Russians called it the Tournament of Shadows. With the British jostling for power and territory on the Indian subcontinent, the Russians scrambled to defend their new southern borders and extend their empire to the Arabian Sea. The games, replete with puppet regimes, double-crossing khans and spies criss-crossing the High Pamirs, dubbed "the Roof of the World," ended late in the 19th century in something close to a draw. Some argue that it has never really ended.

In many ways the antics of the Great Game set the mood and tempo for the relations the Central Asian states still maintain with the rest of the world, and each other today. Egotistical tyrants, the five presidents, solicit favours from foreign powers in exchange for "friendship." Shady deals and tricks, overwrought rhetoric and intelligence – the spying sort – still dominate politics and business here. And the notorious "Central Asian mentality," a much-touted, enigmatic phrase that essentially means a cagey obstinacy to Russians and Westerners, is the persistent excuse for why agreements and understandings keep falling apart. The Central Asians' extreme distrust of outsiders evolved at the time of the Great Game, and the recent wars in Kashmir, Afghanistan and Tajikistan, with their complicated and paradoxical elements, have their roots deep in the quagmire of that era.

Announcing they had plans for a party for my arrival, the "water specialists" all finally left. Then I asked Bozov for a meeting with Rim Guiniyatullin, the head of all components of the Aral Sea project.

Victor had renamed the big boss "Mr. G" after I had repeatedly stumbled over the pronunciation of his name. Mr. G had a reputation as a "wily old devil," a "corrupt tyrant" and an "absolute shit" that had reached mythical proportions. JC Torrion, my contract boss, and Victor had briefed me: Mr. G had been in the Central Asian water business for more than thirty years. He had been part of the Soviet authority that had masterminded the huge expansion of the irrigation systems in the 1970s and '80s; therefore he was partly responsible for the current water disaster. Since the end of the Soviet Union in 1991, he had managed, rather impressively, to establish himself as the region's water powerbroker; no interstate water management decision got through without Mr. G's consent. Or so it seemed, anyway. I was only just getting used to the bluff and hyperbole that accompanied authority figures in Central Asia.

Bozov wagged his head and sighed. *Clearly this Canadian doesn't understand anything*!

"Robert, Robert. Mr. Guiniyatullin is like Uzbek President Islam Karimov. You must wait for *him* to invite *you*. Mr. Guiniyatullin knows you are here and will meet you soon. Don't make him angry because when he's angry we are *both* in big trouble."

Nadir screwed up his face and nodded, adding, "*Big* trouble!"

Then they both shook their heads in overwrought condemnation.

No, this Canadian really doesn't understand a thing!

Victor was frowning and puckering his moustache. I could see that he disapproved of their shtick – Bozov's straight guy to Nadir's mocking slapstick. I could also see by Bozov's scowls that Victor was exasperating him. *This Victor understands too much*!

Bozov demanded something from Nadir, who rummaged through piles of file folders on their muddled desks. Finally he presented me with a single sheet of paper in Russian. Victor looked at it and said it was a CV for a woman named Shakhlo Abdullayeva. Bozov grinned and said she would be my new secretary.

"She is beautiful like a spring flower in the mountains!"

Nadir grinned and added, "And *very* qualified . . . whoa, whoa!"

Victor said something to Bozov in Russian and Bozov snarled back at him. They argued for a couple of minutes until Bozov turned abruptly to his computer and began tapping the keyboard with two plump fingers.

Victor and I stood to leave. Nadir jumped up, gawked and shrugged, and grabbed my hand and shook it.

"Okay, it's finished! You're welcome! See You!" He shook Victor's hand. "See you too! You're welcome too!"

Then Bozov, frowning, reluctantly stood and took my hand in both of his. "*Da svidaniya* [goodbye], Robert!" But he waved Victor away.

By the time the Bolshevik revolution erupted in 1917, resentment of the Russians in Central Asia was strong. During the First World War the tsar had requisitioned men, cattle, cotton and food for the war effort, triggering widespread revolts that the Russians suppressed with bloody crackdowns. To many young Muslims, inspired by their cousins' Young Turk movement in Turkey (Kazakhstan, Kyrgyzstan, Uzbekistan and Turkmenistan all speak Turkic tongues), the Russian Revolution offered hope for a new, more progressive order. Following a brief counter-revolution in Tashkent, the Bolsheviks triumphed and the Soviet era began. But Muslim reformers quickly found that they had been deceived. Moscow ordered Soviet Central Asia, renamed Turkestan, to follow the dictates of its five-year plans. Farms were collectivized, the land managed cooperatively with crops grown and animals reared according to strict decrees, ending nearly 7,000 years of clan farming and nomadic herding. Stalin's era reads like a cattle roundup: millions involuntarily resettled onto collective farms, millions sent to gulags, millions dead of starvation, millions slaughtered for dissent.

Cotton, which had been grown along the Amu Darya and Syr Darya for centuries, was now the focus of agricultural activity in Central Asia. The Aral Sea basin was the only area of the Soviet Union where the crop would grow. It was "cotton fever," spurred on by a nationalistic and competitive desire for self-sufficiency, and five-year plans pushed a continual expansion of irrigated lands to meet ambitious targets. Orchards and vineyards were uprooted, pastureland and fields of fodder plowed under to make way for "white gold." By the outbreak of the Second World War, this fertile region that for thousands of years had produced bumper crops of apricots, apples, melons, grapes, every vegetable imaginable as well as wheat and rice, was unable to feed itself; food had to be imported from mother Russia.

But the worst was yet to come.

Upstairs in our empty office on the eighth floor, Victor stared out the window and brooded.

"Bozov shouldn't talk to me like that," he announced after a long silence. "He should apologize for what he said. He's a very ignorant man."

I had no doubt Bozov was being ignorant. I'd heard him use the word "*Koreanski*" and guessed it was an ethnic slur. Victor was half-Korean, although with his broad shoulders, deep voice and thick moustache his Ukrainian half mostly dominated. Having worked in Asia before, I was used to blatant discrimination and racist comments and jokes. Much of it was just cursory and stupid, like bullies picking on a pint-sized school kid. But there was also no sense of political correctness, no sensitivity that to call attention to someone's ethnicity might be inappropriate. I was regularly being told – with a little laugh – who was Jewish or German or Armenian. But Victor's silent fury indicated there was more to this than just a bigoted insult. Why did Bozov dislike Victor so much? What had happened to sour their relationship?

Despite my questions, Victor wouldn't really offer anything. I was learning that my new counterpart kept things stubbornly to himself. But anyway, I decided, we had bigger problems.

"Victor, how the hell are we supposed to work with this Bozov?" I asked, smiling glumly.

"Robert," answered Victor, his expression remaining stern, "you have to understand Soviet mentality. The lines of authority are very strict. Bozov is only a soldier. It's Mr. G who is the dictator. He orders Bozov to do everything and Bozov, because he's so afraid of Mr. G, will follow orders no matter what. But you see, Bozov can't think clearly for himself. And he's ignorant. That makes him unpredictable." He stroked his moustache for a few seconds and added, "It means Bozov is much more dangerous than he appears."

3   Dancing at the Tea House

The Aral Sea disaster has been blamed on hubris, greed, short-sighted autocratic planning, human folly, cotton, Russian colonialism, the Cold War, inappropriate cost-benefit analyses, a controlled news media, ignorance of the laws of nature, ignorance of scientific warnings, misguided technocratic engineering and patriotic sloganeering. These charges all have degrees of truth. But beyond the finger-pointing, the disaster was ultimately caused by the sort of mad obsession that lays claim to the human conscience when it plots and carries out a murder. The Soviets targeted, condemned and sacrificed the Aral Sea. Then in the post-Soviet era, as it lay dying, a long parade of water specialists, almost all Western, came to study the victim and make expert diagnoses. For every proposed treatment the governments of the five Central Asian states produced an excuse; politics and lack of money assured inaction. And in the end the specialists all sadly shook their heads and departed. It was clear: there was little will to save this disappearing sea.

The massive military build-up during Khrushchev's era set the conditions for the sea's demise. The Cold War created a huge demand for cotton, especially for the manufacture of millions and millions of Red Army uniforms. Added to this was a competitive spirit to surpass the U.S.'s cotton output. Five-year plans dictated ever-higher cotton production targets, new irrigation canals opened up the desert and more and more land went into production. But the expanded production, successful as it was, brought new problems. All the new dams, reservoirs and canals feeding the thirsty cotton crop were bleeding the Amu Darya and Syr Darya. Worse, the short-sighted, sloppy irrigation methods not only wasted water, but also raised groundwater levels and increased soil salinity. In the past cotton had been rotated with other crops, but now it was grown almost exclusively, and this was wrecking soil fertility. Almost as fast as new lands were opened up, older arable lands were being lost.

Thus far the expansion of the irrigation systems had not adversely affected the Aral Sea. In the 1960s tens of thousands of people in the areas around the sea survived comfortably on fishing and agriculture. Each year thousands of kilos of perch, pike, carp and sturgeon were pulled from the sea, canned in local factories and sent all over the

Soviet Union. Cotton, rice and vegetable crops flourished near its shores, watered by local irrigation canals. Millions of muskrat pelts supplied a thriving fur industry. Seaside resorts and pioneer camps lined the sea's sandy shores and during the summers vacationers arrived on direct flights from Moscow.

But in the Soviet capital, Soviet planners had already decided the Aral's fate. Russian scientists had estimated that the sea's level, affected by Eurasian humidity levels, had fluctuated by as much as six metres over the past 4,000 years. Many of the scientists believed these deviations demonstrated that the sea was inconsequential. Such reasoning had begun back in 1908 with the reputable Russian geographer and climatologist A. I. Voekov, who wrote: "The existence of the Aral Sea within its present limits is evidence of our backwardness and our inability to make use of such amounts of flowing water and fertile silt, which the Amu and Syr rivers carry." Calling it a "mistake of nature," he effectively wrote the sea's death warrant.

Shakhlo Abdullayeva was sitting opposite Victor and me in our chilly office. She held a graceful pose, with one shoulder slightly forward and her head proudly raised. She was dressed conservatively in a wine-coloured suit and in a modest Islamic concession, had knotted a matching scarf around her inky-black shoulder-length hair. We reviewed her brief CV. Then she showed us pictures of some Americans she had worked with on an ecological project the year before, funded by the Asian Development Bank. She acted surprised that I didn't know the men in the pictures, who came from Massachusetts and Vermont. She pointed at one, an academiclooking man in his fifties, and said that he e-mailed her frequently.

"He give me Toshiba laptop from America," she boasted. I imagined him completely smitten by her; she was the type of woman who could smite men. More alluring than beautiful, she had clear olive skin, large even white teeth, high cheekbones and an amused glint in her smouldering eyes. As she talked her expressions became more emphatic, her voice shriller and her dark eyes bigger.

We shared the elevator after the interview and she studied herself in the little mirror. She adjusted her hair and then fixed her eyes on me through the reflection.

"If sometimes you don't find me, probably I am here making myself more pretty."

I laughed and after a carefully timed pause, she laughed too, an ironic giggle. I liked her coy determination. She had style and cheek. She was going to be fun.

Followers of geographer Voekov announced that no harmful effects would result from using the water destined for the Aral Sea. They suggested that the extensive irrigation systems would increase the moisture supply to the rivers by intensifying the region's hydrological system. This accepted wisdom provided the rationale for the huge expansion of the irrigation networks in the 1970s and '80s. And it was egged on by the glittering images of Soviet propaganda: dams, canals and cotton fields stretching endlessly into the red desert with cheerful workers singing patriotic songs as they excavated, dammed, tilled and tossed cotton balls into bins stencilled with the red hammer and sickle. But the world's fourth-largest water body had its useful functions: It sustained life and livelihoods. It softened the region's sharp continental climate. Its complex ecosystems were productive and unique. The Aral Sea was a life source, a shimmering splendour in the midst of a stark desert.

Jean-Charles Torrion told me he already knew Shakhlo. On the phone from Paris my boss from BDPA recalled an evening during his negotiations when she'd made *plov*, the Uzbek national dish of fried rice, carrots, raisins and chickpeas, topped with chunks of mutton. Victor and Bozov had been there with him.

"It was at a point when the contract was 'under process,'" he explained. "I thought we were nearly there. It was a sort of celebration. Bozov arranged it and we went to this flat across the street from the project. I think it was Shakhlo's flat." (A month later I would move into it.) "It was set up like an office with desks and a computer. She was very cheerful and chatted in broken English. She seemed very sure of herself. I wondered if she was Bozov's lover." He laughed at this possibility. He was surprised that I was planning to hire her, but he didn't raise any objections. "Did Bozov recommend her?"

I said that he had, adding, "Hiring her might get things off to a good start."

JC laughed again. "Have you met with Mr. G?"

"Not yet," I admitted. "Bozie's keeping me to himself at the moment. But I'm working at it."

25

The experts call the Aral Sea's slow death "a creeping environmental disaster," although 40 years in the sea's 35,000-year life is barely a blink. Milestones have marked its downfall. The first was reached around 1975 when production peaked; it became impossible to grow any more cotton in the Aral Sea basin as all available land was already in production. And as nearly half of this land was dangerously saline, overall production began to decline. Then, in 1982, not one drop of Amu Darya water reached the Aral Sea for the first time; the cotton fields and the sun had sucked it all up. For two decades the river's flow had been decreasing, causing the sea's shoreline to recede. Each year the beaches grew longer and longer and the salinity of the water increased. By the early 1980s the fishing boats were permanently beached and the Russian holiday camps long gone. Most of the fish had also disappeared. The sea's volume had shrunk by almost 50 percent. And upriver, the second-rate irrigation systems were destroying the natural drainage systems, allowing discharged water to pollute rivers and groundwater. But worst of all, the heavy use of fertilizers and chemicals to grow cotton was poisoning the land, the water and the people.

A different kind of milestone was reached in 1984. Moscow, using new spy satellites, discovered that many of the so-called cotton fields in Uzbekistan were in fact only tracks of desert. Uzbek officials had been grossly inflating annual production figures and tricking the Soviet apparat into paying them millions of rubles for nothing. And it had been going on for years. The gig was finally up and the Soviets fired thousands of officials. But they never charged the ringleader, Sharaf Rashidov. In fact, at Uzbek independence, President Karimov recast Rashidov as a national hero, along with the despot Tamerlane. The spirit of the Great Game endures.

At the same time Soviet planners also realized that gross water mismanagement and defective irrigation technology were resulting in growing economic losses. In response they trumpeted a new scheme that would solve all their problems. They proposed reversing the flow of Siberia's Ob and Irtysh Rivers south into the Aral Sea. The "Project of the Century" was megaproject madness. Apart from the ecological damage, which the Soviets ignored, the scale and costs were titanic. Environmentalists were alarmed: *Trump a disaster with an even bigger disaster*! (Later they would learn that the water was not really meant for the Aral at all but for the further expansion of irrigated lands in the desert.) The scheme turned out to be one of the last fanatical rolls of the dice of the cocky Brezhnev era.

26

"Robert, you are a very smart Canadian!" Bozov was thrilled with the news. But Victor, although he agreed that Shakhlo was the best of our five candidates, was not.

"Her voice is too shrill," he said firmly. "It won't sound nice when she answers the phone."

Shakhlo began scrounging desks and chairs from the project storeroom and arranging our three small offices. Victor and I bought a safe and she spent a long time learning the combination, turning the dial right and left and demanding I watch her to see if she was doing it right. She told me she had found some textured wallpaper and arranged for some decorators to hang it. She would buy curtains. When I asked her how much all this would cost, she studied me for a few lingering seconds.

"Of course I get good price. You don't trust me?" She shook her hair free – the head scarf was gone – and looked at me as if I could not doubt her.

One afternoon Nadir came upstairs to my office and announced, "Mr. Bozov is waiting for you! Please come only yourself. Do not bring Mr. Tsoy!" Victor was out of the office and Nadir confided, "Mr. Bozov does not like Mr. Tsoy's interpretation. He says it's not correct! So come now, please. *I* am your interpreter."

We both laughed at this, knowing it was Nadir's translations that were terrible.

The rooms below us hadn't paid their heating bill. It was late January and with the radiators off our offices were frigid. I mentioned this to Bozov, who grinned. He said he was surprised, but I could see that he wasn't. He suggested I buy electric heaters. But I said Victor had checked the outlets and found our office was all on one circuit.

"If we plug in heaters, we'll blow the circuit."

"Add more circuits," said Bozov through Nadir. They swapped grins.

JC had told me that Mr. G had pressured him into our office lease. The rent was excessive and I suspected Mr. G and his lieutenants were pocketing the rent money – so far they'd refused to provide us with a proper receipt. "This seemed to be something Mr. G really wanted," JC had rationalized. He had given in to Mr. G on this, just as I had hired Shakhlo largely for Bozov.

Bozov shrugged. "Robert! This is not your problem. This is a problem for Victor Tsoy!" He and Nadir laughed. Then he invited me out to a restaurant that evening. "To celebrate your arrival! Everything is arranged. Bring Shakhlo, but don't bring Victor."

By 1986, with Gorbachev's perestroika and glasnost allowing the airing of dirty laundry, the Soviet media declared the "Aral Sea situation" a disaster. Central Asian journalists and poets, environmentalists, scientists and engineers, newly hatched NGOs and suddenly revisionist government officials formed a public front to save it. The disaster acquired new sensational labels – "the Aral Sea catastrophe," "the quiet Chernobyl" – and by the mid-1990s it had taken on status and gravity as one of *the* great environmental issues of our times: the Brazilian rainforest, Chernobyl, the ozone hole, global warming, the Aral Sea. Its name pushed a water distress button, much as Chernobyl had pressed one for nuclear power. The Western media picked up the story: The Dead Sea; Saving the Last Drop; Despair Looks Like a Sea That Died; Eternal Winter: Lessons of the Aral Sea Disaster. By the end of the century the world knew all about it.

But did they care? This disaster stood apart from other global environmental issues in two critical ways. First, it was the 20th century's biggest water crisis, the first time water had become a scarce resource on a massive scale. Second, it was taking place in that black hole in the atlas called Central Asia. Who in the West could place Uzbekistan on a map? Or knew the capital of Kazakhstan? CNN and BBC World ignored the region, an area the size of Australia, in their global weather reports. To most Westerners, Central Asia was an isolated wasteland of desert and mountain and sheepherders that didn't matter. (Unless, like Afghanistan, it burst into their homes via their TV screens.) The disappearing Aral Sea was an aloof disaster, one that didn't touch their hearts and minds.

In 1988 the Communist Party of the Soviet Union admitted for the first time that a serious mistake had been made in the use of land and water resources in the Aral Sea basin. They took no new measures to alleviate the calamity but did officially kill the Siberian rivers diversion plan, the "Project of the Century," although the mad scheme keeps refusing to die. They also invited the World Bank and the United Nations to send in their experts to investigate the disaster and find solutions. In 1991 the Supreme Soviet passed a resolution outlining a huge land reclamation project that they touted would save the Aral Sea. But it was never carried out. Weeks later the Soviet Union was history and the Central Asian states unexpectedly sovereign. And Russia, in her new laissez-faire mood, backed away and shirked any responsibility for its seventy years of gross water mismanagement.

On the edge of Babur Park, the *chaikhana* ("teahouse") looked like the interior of a sultan's tent. Dark red canopies swept down from the ceiling to create intimate niches for the tables tucked along the walls where couples leaned close, their faces almost touching. Blood-red Turkmen carpets covered the floor, overlapping and forming humps in places, but cleared from the centre of the room for dancing. A deejay, propped among some amps at the back of the room, was playing Russian pop songs.

Our long table was in the middle of the room. There were about twenty seats, half of them already taken by members of the Kazakh public awareness team or by Component A, the Kazakh component. Bozov's cronies at the project all seemed to be Kazakhs. I sat with Shakhlo to my right and Bozov to my left. My jigger of vodka was filled and refilled. There were introductions each time a new face appeared.

When the table was full, Bozov rose and made an introductory toast. Shakhlo provided me with a rough translation:

"Robert is finally here. Congratulations! Soon he will decide which country in Central Asia has the most beautiful women!" Loud cheers. "Now that he is here, our public awareness work can be completed. We have already done good work and we have been waiting a long time for him to come to tell us how excellent it is!"

Everyone cheered again and knocked back their glasses.

While Bozov was having some fun with me, it was also clear that I was a threat. He didn't trust me. His jollity was mocking and condescending. I was a Western consultant and he was an old Communist. To Bozov the Cold War had never ended.

The Kazakhs tore up nan, round flatbread sprinkled with caraway seeds, and passed it around. Beetroot, carrot and cabbage salads arrived, followed by *plov*. The meal was interrupted with repetitive welcoming toasts and our vodkas topped up over and over again.

Towards the end of the meal I told Bozov I would make a toast. I stood up and floated above a sea of grinning Central Asian faces. I was completely smashed.

"I am the happiest man in Central Asia because I'm finally here!" That was true; despite the undercurrent of animosity I was determined to rise to the challenge. Shakhlo translated and there were cheers. "I know what a pleasure it will be to work with all of you on a project as important as this one. The world is watching, especially the World Bank, and we will show them that we can persuade the people of Central Asia to use less water and save the Aral Sea. To our successful work together!"

More cheers and everyone knocked back their vodkas.

"Robert!" Bozov leaned close to me and wrapped his arm over my shoulders. "It was good, but you shouldn't mention the World Bank. We don't need them." His plump face leered at me as Shakhlo translated this. "And now you know Central Asian hospitality. Look! So many beautiful women, like spring flowers in the Kyrgyz mountains!"

Shakhlo looked coyly at Bozov and laughed.

An Uzbek crooner was wailing a popular Uzbek song and Shakhlo lured me into a dance. She showed me the moves – raised arms fluttering beside your head and hips gyrating. It was the swim and the hula combined with a belly dance. I tried to follow her and soon everyone was laughing at my staggering moves. It was make-fun-of-the-expat, but I'd been through the routine before, many times, in Mongolia. And I was too drunk to care. The Kazakhs joined us and we wiggled and flapped and yowled around the room. Shakhlo was swirling and jiggling with the grace of a professional belly dancer. She kept eyeing me, pulling me back into her dance.

The next morning Shakhlo told me I'd left my jacket at the restaurant and the driver had had to go back for it. She also said that one of the women in the restaurant wanted to see me again. She gave me a coy look and then her bewitching smile. She was teasing.

I saw her dancing again. She had exuded sexuality in the way she'd moved, the way she'd held my gaze, the way she'd twisted in her dress and shaken back her dark hair. She'd led with poise and confidence. But there was something conniving about her. This mother of two, a woman verging on middle age, was too determined, and a little devious.

"You dance with her, again and again," she said. "She ask me who you are and I tell her you are *Canadian*. She is very pleased."

"But Shakhlo, I can't remember any other woman in the restaurant. Only you."

Following the collapse of the Soviet Union, the Aral Sea basin was sliced up into five adolescent states and its scarce water resources were suddenly a source of fierce competition. The five Communist Party leaders, instantly recast as "democratic" presidents, held onto power by eliminating their opposition. Almost broke and desperate to bring in hard currency, they turned again to cotton, still their biggest cash export. They demanded even more be produced, ignoring all the experts' warnings.

And so it was. Water tapped from the Amu Darya and Syr Darya continued to gush through the 32,000 kilometres of canals, 45 dams, and more than 80 reservoirs. It poured out half-broken cisterns, slopped onto furrowed fields and evaporated in the blistering heat. The poor drainage leeched so much salt to the surface that it formed crests like windblown snowbanks in early spring. More chemical pesticides poisoned the land. But now all five presidents knew full well the consequences of their actions, and they could no longer blame the Soviet Union.

By the end of the 20th century most of what was once the Aral Sea was now the Aralkum, a 38,000-square-kilometre man-made desert. The remaining sea, shrunk into the "Big" and "Little" Aral Seas, held about twenty percent of its 1960 volume. The wind whipped up eddies of toxic dust, a mix of cotton pesticides, heavy metals and salt that whirled around the bleak settlements on the edge of the new desert. In rundown hospitals in Karakalpak Autonomous Republic in Uzbekistan, Kyzyl-Orda province of Kazakhstan and Tashauz province of Turkmenistan, patients wasted away with liver and kidney diseases, TB and other chronic respiratory illnesses. There were high rates of birth abnormalities and many children suffered from asthma and neurological diseases. And the ecosystems had mostly shut down. Twenty species of Aral fish and most of the bird life were gone. The cotton and rice crops were failing because they couldn't get enough water, and the water they did get was too saline. The weather, too, had changed. Winter was now a month longer and a few degrees colder, the summers even hotter, and the number of rainless days was increasing.

Leaving the office one evening, Victor and I stood in front of the elevator, listening to it slowly creak between the floors. When it fell silent we heard a heavy rasping coming from down the hall and turned to look.

A huge bald man in dark glasses was looming towards us. The jacket of his baggy black suit hung open and his dark tie, pulled violently loose at the neck, was resting on his enormous belly. He lumbered up to us and we stepped back to let him through. He was scowling and a cigarette dangled from his lips. I gazed at him in awe, thinking I'd never seen anyone look so sinister.

I heard Victor introducing us, but was too astounded to catch his name.

"*Guten Tag*," he growled as if we were German. He was glaring into my face. I extended my hand but he ignored it. The ash on the tip of his cigarette was on the verge of crashing to the floor, but somehow it didn't. My eyes locked onto it. Victor was saying something else in Russian, trying to connect us. Then I felt the stranger take my outstretched hand. His grip was surprisingly limp.

"*Zdrastvuyte*," I managed, the formal Russian hello. We stared at each other for a long speechless minute. I couldn't think of anything to say, in German, Russian or English. And he felt no obligation.

The elevator began to squeak again and he turned away as the doors finally opened. He waddled in, taking up the entire space. Then he turned around and continued scowling at us. By rights the elevator should have been ours, I thought with a silent laugh, as we were there first. Or maybe we could have squeezed in beside him. But we didn't dare make a move. The doors closed with the ash of his cigarette still dangling.

"Who was that?" I asked, looking at Victor. I was still in shock.

"Mr. Guiniyatullin." He smiled. "The notorious Mr. G."

"I feel like I just met the devil."

Victor puckered his moustache several times.

"Very likely you did."

4   A Gift from Allah

Standing at the head of the long table, Bakhtiyar Nazarov poured a little tea into a bowl and returned it twice to the pot. Then he passed it to me in his right hand while placing his left hand over his heart. The cup only held a few sips but I knew this polite formality would allow him to offer to fill up my cup again soon.

A man of custom and ceremony, Nazarov was the leader of the Uzbek public awareness team. We were in the Uzbek Academy of Sciences, where he was also deputy director. He had invited Victor and me to his office to discuss our future work with his team of journalists, teachers and water specialists. Like the four other teams – the Turkmen team had yet to form – his had been producing information and disseminating it through the Uzbek media for almost a year.

His office smelled of well-steeped tea and the old leather volumes that filled two rambling bookcases. A large potted rubber tree grew indulgently in a corner, its limbs reaching towards the tall windows, half-covered in faded gold drapes, that gave the room a tawdry elegance. The view across Gogol Street was of a sand-coloured Stalinist building similar to the one we were in.

Speaking with a politician's inflection, Nazarov reviewed his team's activities. One of them was a contest for every schoolboy in Uzbekistan: After learning about their water traditions from their fathers and grandfathers, they wrote about them in compositions. The best ones were going to be published in a booklet that he said would be distributed all around the country.

"They learn that water is a gift from Allah," he declared, smiling warmly. A few gold teeth glinted.

Nazarov cupped the teapot in both hands and we returned our bowls and watched him repeat his meticulous pouring.

"It was a good first meeting," said Victor as we drove back to our office. "Nazarov is well connected. He used to be one of President Islam Karimov's advisers. He's part of his inner circle. That's why he got the job." He smiled sarcastically.

The Uzbek president was a dictator and his *militsiya* clamped down hard on dissidents and disregarded human rights. Tashkent felt repressive; the military was watching on every corner. The cost of living was escalating and only Karimov's connected friends seemed to be benefiting from the stifled economy.

"But his team's materials are boring," continued Victor. A few days before Bozov had presented us with a folder stuffed with newspaper articles, booklets and pamphlets as well as several videos. Victor had gone through them. "They blame the Soviet Union for the demise of the Aral Sea and provide the opinions of local water specialists on rational water use. They praise the Aral Sea project and claim their specialists have the answers to solve the terrible problems the Soviets created. But the public is not stupid. They won't believe any of this. And anyway, they won't care. They're used to this kind of Soviet propaganda."

"Do you think Nazarov will cooperate with us?" I asked.

"I think so," answered Victor. "Do you know why? Because he didn't mention Bozov. I think Nazarov doesn't respect him. That's why I think he'll be on our side."

"I didn't know we were taking sides," I said.

Bozov kept a Russian version of our contract, dog-eared and doodled on, front and centre on his cluttered desk. It was his bible and he clung to it. Whenever he thought I wasn't obeying orders he picked it up and thrust it at me, bellowing, "You are not following the terms of reference!" with Nadir echoing his words. But BDPA's actual proposal was nowhere in sight and I found out that it had never even been translated into Russian. Yet even after I had this done, Bozov ignored the document. Still I used it to put together an action plan – a standard procedure on international projects like this – but he objected to every suggestion I made. (Over the course of my year in Tashkent he refused to approve any of my action plans.)

BDPA's proposal advocated a series of campaigns aimed at the most wasteful users of water in Central Asia. Instead of the public awareness teams sending out general information that lamented the fate of the Aral Sea and organizing competitions that educated schoolboys to the preciousness of Allah's great gift, specific messages would target farmers, irrigation system managers and key officials who were responsible for the five states' excessive water use. The goal was to trigger these groups to change their behaviour and reduce overall water consumption. The approach included four elements: developing an overall strategy to coordinate activities; a lobbying component to ensure the five governments were on side; running a series of training workshops for the five teams; and providing PR that promoted the issue to Western donors and media. It was my job to turn this approach into reality. But with Bozov stonewalling my attempts to get things going, I was starting to look for ways around him. Nazarov was my first potential ally. But as Victor saw it, what we were really doing was taking sides against our boss.

On a crisp Sunday in early February, Victor and I headed up into the Chatkal Mountains, an arm of the Tian Shan range that reaches across the Kyrgyz border to touch the suburbs of Tashkent. Victor had worked as a tour guide and was in his element, enlightening me on the subtleties of Uzbek life. We rode all day in his hobbit-sized Daewoo Tico, through the Chirchik Valley and circumnavigating the Charvak Reservoir to view the upper reaches of the nefarious irrigation systems. The chimes hanging from his rear-view mirror played softly every time we hit a bump, like a child endlessly opening and closing a music box.

Along the highway dozens of middle-aged women wrapped in quilted coats and kerchiefs were selling apples, which they'd polished and arranged according to colour in leaning crates – vivid reds, yellows and greens set against the bleak winter landscape. We stopped and bought some.

"The other day Mr. G stopped me in the hall and asked me why you hadn't rented any of the flats you were shown," said Victor as we crunched our apples and gazed at the snowy mountains. The previous week a project officer had taken me to see several apartments, recommending them for BDPA's team of consultants. "I told him that I was helping you arrange housing. But he didn't like that."

None of them had been right. They were either too far from the office or the wrong size. The flat that was supposed to be for me was huge and enormously overpriced. I figured they all belonged to Aral Sea project staff and that they were just trying to bilk BDPA.

"He also asked me to report directly to him on our work," continued Victor. "But I told him that I work only for BDPA." He glanced at me and twitched his big moustache. "I'm not going to be Mr. G's spy," he said, tossing away his apple core. "Of course he was angry. He said that if I didn't cooperate I wouldn't work on this project. But I told him that my job was Robert's responsibility, not his."

35

I was starting to see why Mr. G and Bozov hated Victor so much. He was gutsy and defiant. He ran his own NGO and NGOs threatened their Soviet-style authority. His Korean-Ukrainian background marginalized him from the five indigenous groups that had dominated politics in Central Asia since the end of the Soviet Union. And as much as I admired his courage I was starting to wonder if his obstinacy was going to keep us from ever working with our bosses.

"Maybe you should have said yes," I said with a smile. "After all, what have we got to hide? You could be a double agent. Report back to me on what Mr. G is doing."

"Robert, you must take this more seriously. Mr. G is looking for ways to control us, especially you. He's very suspicious of foreigners. You don't know what he's capable of doing. Spying and blackmail are normal activities in Uzbekistan."

"You make it sound like the Great Game all over again."

Victor eyed me dolefully. "That's just what it's like."

Victor went on to explain Rim Guiniyatullin's background. Mr. G had been a major figure in water management in Central Asia for close to thirty years. He'd risen through the layers of Soviet bureaucracy at the agency that executed the huge expansion of the irrigation network, eventually becoming one of the masterminds behind the emptying of the Aral Sea. With the demise of the Soviet Union, he'd shrewdly played off his enemies and continued to direct water management decisions through various positions he'd held at the Uzbek irrigation authority. He had unique credentials: An ethnic Tartar, he spoke fluent Uzbek, which gave him Uzbek status, and the Uzbeks have always been the region's most powerful ethnic group. An ex-Soviet water expert with an autocratic style, he was both feared and respected and his staying power contributed to a sense of stability; with Mr. G in control it seemed that things were continuing just as they had in good old Soviet times. The Water and Environmental Management Project for the Aral Sea Basin got underway in 1997, operating under the umbrella of the International Fund for Saving the Aral Sea (IFAS), an agency that Rim Guiniyatullin happened to run. But with power sharing and conflicts of interest holding little sway in Central Asia, he also landed the job as head of all components of the project. These two top jobs allowed him to consolidate power and become the region's water management tsar.

"Maybe we won't have to contend with Mr. G for much longer," Isuggested. "Maybe a coup in Ashkhabad will solve things for us."

We'd been hearing rumours of discontent. Turkmenistan, currently chairing IFAS, strongly resented Guiniyatullin's domination of the project, and in Central Asia chairing a committee meant autocratic control. There was almost no sense of democracy. Bozov had told us that the Turkmens wanted to move the project from Tashkent to Ashkhabad, the Turkmen capital, adding ominously that Mr. G was not going to attend the upcoming IFAS meeting in Ashkhabad, a move that would stalemate decision-making.

"Robert," said Victor disparagingly, "even if the Turkmens change Mr. G, they'll only replace one dictator with another."

At the Chemgan ski resort there was a long lineup at the lift, but Victor spotted a former colleague near the front of the queue and we cut in without objections from the crowd, all Uzbek Russians decked out in fashionable skiwear. Victor's friend was a journalist and he wanted to know what was going on in Canada.

"Hockey? Anything else?" he asked snidely. "What is there to do in Canada anyway?" Then he wanted to know what kind of public awareness work we would be doing. "Giving out information on the Aral Sea catastrophe? Nobody here cares about the Aral Sea. Why should we? We have much bigger problems than the Aral Sea." He was speaking loudly and there was some laughter around us. One man about to get on the chairlift shouted, "We didn't create this problem!" There were cheers.

"You see," explained Victor as we rode up in the lift, "the Russians here are bitter. They resent the Uzbeks. They have lost their power and now they blame the Uzbeks for all their problems. But everyone in the country must take responsibility for the Aral Sea catastrophe. Our job is to convince the public that what had happened to the Aral Sea was not only an environmental disaster but a reprehensible act, and that the rehabilitation of the sea is necessary not only to end the degradation but to absolve the wrongdoing."

"Irrigated farming began on the Amu Darya and Syr Darya rivers in the sixth century BCE. This was then Sogdiana, and the remains of their ancient irrigation systems are still evident today. For more than a thousand years, traditional approaches to water use were passed down, from generation to generation. The people respected water. They used it frugally. The code of regulations in Muslim law, the *shariah*, ensured that water was not wasted. Extravagant use of water was considered a sin. The people observed regulated patterns of cultivating land and used water in the most reverent way. At that time water was considered to have the same value as gold. Every drop was sacred. When a son was born, his father planted a poplar near a canal. The water from the canal nurtured this tree and gave it life. When it was fully grown, they cut it down and used it to build the family home. And every spring, the people of each village gathered together to clean the canals and reservoirs and repair the roads and bridges. This was called a *khashar*."

We were sitting in a big office listening respectfully to Abdurakhim Jalolov, a middle-aged Uzbek with pomaded hair and an unctuous smile. Jalolov was the Uzbek national coordinator for the Aral Sea project. Each of the five participating states had a national coordinator for the project; it was their job to stamp their government's approval on anything undertaken by any of the project's five components. As the job was highly political, they each held a high position in their governments. Jalolov was the first deputy minister of agriculture and water for the government of Uzbekistan.

His speech went on for nearly an hour. He warned that Allah was taking revenge for the brutal interference in natural events, for the irrational use of natural resources in the Aral Sea basin. The Soviet Union had expanded irrigation systems and instilled a careless attitude towards their precious water. Such wastefulness had never been allowed in Central Asia before Soviet times, he claimed. The Aral Sea was disappearing because Uzbeks had lost their respect for water. They had forgotten their traditions.

"The *shariah* says, 'If there is water, there is life and prosperity. If there is no water, there is hunger, death and migration.' This code must be enforced. Do we think about the harmful effects that a new canal will have on the population today? The *shariah* says we should not build a canal or a water system if it will have a negative effect on any person. We must follow these rules again. We cannot reach agreements with our neighbours on how our water resources should be shared. We ignore the fact that the future of the region depends on our rational and responsible use of water. The *shariah* states that all water users must bear the cost of improving and cleaning our waterways. If a dispute arises between neighbours over water, it should be settled on the basis that water is allocated pro rata in each area. The farmer on the upper reaches of a river can only use the water allocated to him by mutual consent of his water co-users downstream. He cannot use more. We have forgotten these promises. Today we are wasting our precious gift from Allah. We don't respect our neighbours. We take as much water as we want. We are spoiling our lands and squandering our water. Our children will suffer for our greedy behaviour.

"I suggest to you, *monsieur*, that what your public awareness activities must do is to revive these traditions." I looked up and he was glowering at me. "The people of Central Asia are not wasteful, greedy people. But they are sleeping. You must awaken in them their traditions. You must remind them of the *shariah* and the Qur'an. 'Eat, drink,' said Prophet Mohammed, 'but do not waste.'"

"Do you think he's really serious?" I asked Victor after the meeting.

We stood leaning over a railing, looking at the Ankhor Canal. The Ankhor is one of several canals that crisscrossed Tashkent, channelling water out to the surrounding cotton fields. The water was glacial green. (A few months later I would swim in it and discover it to be surprisingly clean and refreshing.) It flowed past the Ministry of Agriculture and Water building, painted an appropriate broccoli-green, on Navoi Street, under a bridge and on to Mustaqillik Maydoni (Independence Square), the huge parade grounds called Lenin Square in Soviet times.

"You see," said Victor analytically, "reviving traditional approaches to water conservation is a safe way to talk about the problem. It avoids taking any real responsibility. Jalolov sees public awareness as propaganda that tells the people that the Uzbek government is not to blame for this catastrophe."

We watched a branch swirl by in the current. It got caught against one of the bridge supports, lingering for a few seconds before being sucked downstream.

"If the Uzbek government was serious about cutting back on water use in agriculture," continued Victor, "they'd change their policies. Then you'd see much more water -saving than we can ever generate with our public awareness activities."

Shakhlo was standing on a chair measuring the windows of the office I shared with Victor.

"Soon here are curtains. Just like you tell me." She stepped off the chair, straightened her dress and looked seriously at me. "I try very hard to make everything good for you."

"Thank you, Shakhlo. I appreciate it."

"I hope you will like new curtains."

"I hope so too." But her expression suggested that I was difficult to please. "I'm sure I will," I added with a smile.

I gazed out the naked window. The view looked north across Tashkent. Directly below was wide Abdulla Kodiry Street and beyond it, Microdistrict 13, a collection of identical five-storey apartment blocks. Filling the gaps between them were children's playgrounds, dormant gardens and stands of leafless trees.

"You look at flat today?" asked Shakhlo. "I show you after work."

Shakhlo was pushing me to rent a flat in one of those five-storey blocks, the one where she'd served *plov* to JC Torrion. She said it belonged to a friend who lived in Moscow. Every day she was reminding me about it and by her insistence I knew she would profit if I took it. I'd been staying in the flat reserved for JC and I needed a home. Yet I put off going to see it. I couldn't quite face living in one of those cheerless buildings below. I had lived in Soviet blocks like these before, in Ulaanbaator and Bishkek, when I'd worked in Mongolia and Kyrgyzstan. The apartments were always much better inside than they appeared outside, but the address, something like Microdistrict 13, Building 17, Entrance B, Flat 27, disheartened me. These districts were dreary mazes.

Beyond Microdistrict 13 lay another Tashkent, dense neighbourhoods of Uzbek-style houses and tree-lined alleys. Called *mahallahs*, these areas had been private even in Soviet times. Two-metre-high walls enclosed each house and the rooms opened onto garden courtyards where grapevines hung over trellises and fruit trees shaded fountains and pools. Many had evolved over the years: a second floor added, a panel of engraved wood set above the front door, tinted windows installed so no one could see in. Some of them turned into palaces with huge garages, marble facades and swimming pools.

Recently Victor and I had visited the offices of *Médecins Sans Frontières* (Doctors Without Borders) in a house in one of these *mahallahs*; MSF was undertaking public awareness work in destitute areas near the Aral Sea. After our meeting I'd said to him, "Why can't I live in a wonderful house like this? Why do I have to live in a dismal Soviet flat?" But I knew that these houses were bigger than I needed and the rents three times as much as a flat. Then I thought, why not rent one for our visiting consultants? If it was large enough it could accommodate all five of them. We could also hold our workshops there.

Shakhlo was still looking at me.

"Okay, let's go see the flat after work today."

She brightened. "Of course you will love it!"

"Why is Mr. Ferguson here?" demanded Valentina Kasymova, the leader of the Kyrgyz public awareness team. "Why do we need foreign specialists? We are specialists in public awareness! Nobody told us foreign specialists would come to evaluate our work."

Bozov's "spring flower from the mountains," "the best public awareness specialist in Central Asia," sat down and looked put out.

We were sitting around a big table in the project's meeting room. Bozov had ordered all four team leaders to Tashkent. On one side of the table sat Bozov, Nadir, Victor and me. Opposite us were Dauletyar Bayalimov, the Kazakh team leader, a balding man in his fifties with squinty eyes and elfish ears; Valentina Kasymova, middle-aged, matronly and humourless; Talbak Salimov, the handsome curly-haired leader of the Tajik team; and Bakhtiyar Nazarov, the Uzbek team leader I'd sipped tea with a week before. Watching from the wall were the regions' five presidents, looking serious and presidential, photographed at the signing of the IFAS agreement.

The Kazakh leader stood up.

"We are working well now!" shouted Bayalimov, banging his fist on the table. "Let us all agree that the best water experts in the world are right here in Central Asia!" He grinned impishly.

Next Nazarov, the politician, rose and surveyed the room.

"I have already met with Mr. Ferguson," he said, "who I understand is a water specialist from Canada." He acknowledged me with an assured smile. "I know that he is a very good man, but I do not know why he is here. Mr. Guiniyatullin has said that we are all specialists and that we know how to do our jobs. We have been working successfully already for one year!"

Talbak Salimov was next, and the Tajik leader sounded more hopeful. He at least seemed to understand that I was here to help them. But then Bayalimov banged the table again and shouted something that made everyone laugh. Nadir looked at me with feigned shock. Bozov was grinning to himself and doodling on a piece of paper.

I was thinking: They've strapped me with four old apparatchiks, defensive and stubborn dinosaurs stuck in the Soviet era. They completely distrust Westerners. They'll be closed to any of our suggestions, just as Bozov is. I won't get anywhere with them.

It had not been like this in Mongolia. I had spent two years there training NGOs and government agencies how to run their own public awareness campaigns. Admittedly they were small-scale, but many of them had worked. They were really pilot projects, a series of activities complemented with brochures and posters and a media blitz. Each one

41

identified an environmental issue – such as an endangered species, deforestation or spring brush fires – and came up with messages, incorporated into the materials, that targeted groups that could do something about it. Not all of the campaigns were successful, but we ran more than a hundred in two years and the most effective ones became models for bigger ones that, in theory at least, the NGOs and government agencies could run themselves. Money became a problem – the Dutch funding ran out and despite everyone's agreement that the project was a success, no other donors were forthcoming – but there was never a lack of enthusiasm or suspicion about our intentions.

I told them all this, explaining who I was – not a water expert – my experience in Mongolia and my wish to share my ideas and develop new ones with them in Central Asia. I told them that the problems in Mongolia were tiny compared to the scale of what we were facing here, but that some of the methods were similar. I proposed trying a number of pilot activities, with farmers and water managers and involving government officials, to see how successful they might be. As in Mongolia, we could then expand the ones that worked into major campaigns. We were lucky; we had lots of money. But time was running out and we had lots of work to do. I told them I was thrilled to be part of this project, how unique it was, how the rest of the world was watching to see how we fared. "Let's show them we can make public awareness work," I said, ending my pitch.

There was a long silence broken only by Victor, who whispered, "I hope you will be able to realize your dream."

Then, "Who will pay for my flight from Bishkek to Tashkent?" It was Valentina again. "Who will pay for my hotel? Is Mr. Ferguson paying for them?"

"I think it's time to visit the World Bank," I said to Victor.

The exterior of the World Bank's offices on Academician Suleymanov Street in central Tashkent was pure Stalinist, but the interior had been renovated into stock Western offices. The furniture was solid and business-like. Wall-to-wall beige carpet covered the floor. Lacklustre watercolours of Central Asian sights hung on the walls. Only the half-closed blinds, diffusing the afternoon sunlight into ribbons of shadow, softened the corporate atmosphere.

Officially the Aral Sea project was not a World Bank project as its US$250 million came from the Global Environment Facility, an international fund that supports environmental projects but allocates their management to other agencies, such as the World Bank or the United Nations Development Program. The World Bank was responsible for monitoring the project, allocating the funds according to their procedures and advising Mr. G and the five component directors on its decisions. But the independent-minded bosses of the project viewed the World Bank more as passive observers. "We don't need them," Bozov had boasted that drunken night at the *chaikhana* in Babur Park.

Anatoly Krutov, the World Bank officer in charge of the Aral Sea project, was about forty. An Uzbek-born Russian, he stood just over five feet tall and looked lost behind his big desk. He had a pert, upturned nose and a sunny expression that he dropped like a mask whenever he took the phone or someone poked their head into his office to ask him something.

"Anatoly, did the team leaders even know I was coming? They've just told me they don't want to have anything to do with me. They don't understand why I'm here and think I'm a water specialist. I tried to explain my role, that I was a *communications* specialist and that I was going to help them with their public awareness campaigns. But they refused to listen. And Mr. Bozov is playing along. What's going on?"

"Robert, they're scared of you! They think you will criticize their work." Anatoly laughed. He held a master's degree from a U.S. university and spoke excellent English. "You must talk to Mr. Guiniyatullin. He will help you."

"I've been trying. But his secretary won't make an appointment for me. And Mr. Bozov won't let me anywhere near him. He says I must wait for Mr. Guiniyatullin to invite me."

"Sometimes Mr. Bozov can be a little difficult," admitted Anatoly with a smirk, as if everyone knew the problem with Bozov. "But really he takes orders very well. I'll call him and tell him he must arrange a meeting with you and Mr. Guiniyatullin. Don't worry! You're only experiencing some of the infamous Central Asian mentality. You will get used to it!"

Anatoly seemed to understand well the challenges I was up against. He will help me out here, I thought. He's on my side.

"Robert," he said as I was leaving, "You and your team are their *first* foreign specialists. They haven't worked with the enemy before. They are testing you, trying to figure you out. You and BDPA are the *guinea pigs*." He laughed and looked pleased with his idiom.

Robert Ferguson

But despite Anatoly's promise, Bozov still refused to arrange my meeting with Mr. G.

"You shouldn't go crying to the World Bank with your problems," said Bozov the next day with a sneer. "Krutov is not your boss. He won't be able to help you."

5   The Famous Blue Cupolas

Muddy water filled the ditches on both sides of the M-39, the modern Silk Road. It slopped out of cracks in the open concrete cisterns that stretched in straight lines from the highway out to the distant fields, many of them flooded. Planting was two months off but already the "washing of the fields," as it was known locally, was well underway.

Victor kept his stern eyes fixed straight ahead and his lips puckered under his bushy moustache. He looked grimly obsessed. Driving, like everything he undertook, was a serious business. We were on our way to Samarkand, the ancient Silk Road city founded in the fifth century BCE, about 300 kilometres southwest of Tashkent. Victor had some "NGO business" to attend to and I was happy to be a tourist and take a break from my ordeals with Bozov.

A few Mercedes, Korean Daewoos and Russian Volgas (a sort of Soviet version of a 1970s Buick) sped past. An Iranian transport overtook us, bound for some mysterious Persian city far beyond Samarkand and Bukhara and the Karakum Desert of Turkmenistan. The Farsi swirls and dashes on its trailer turned the familiar image of a semi rumbling and hissing down an open highway into something exotic.

The dormant land half-submerged in murky water was reminding me of the gravity of our mission. Water use in this region was the heaviest in the world, with demand for the flows of the Amu Darya and Syr Darya already exceeding supply by 25 percent. The massive regional irrigation system that the Soviets had created was coming apart at the seams, squandering almost as much water as it delivered to the fields. And the staggering waste was getting worse. The five former Soviet states, beset by economic and political woes, were spending on average less than US$10 a hectare on maintenance – a drop from $25 a hectare in Soviet times, when the system was in much better shape. And with the population growing 15 percent a decade, the governments had to come up with better ways of growing food to feed all the extra bodies. It all added up to a water crisis looming not too far down the road.

Component B was intended to be part campaign to stop the wasteful behaviour of heavy water users and part exercise in PR. Its formulators at World Bank headquarters in Washington had envisioned a two-pronged approach: one, to build public awareness of the rapidly disappearing Aral Sea, thereby helping to defuse the impending water crisis; and two, to get the people behind the Aral Sea project. It was an experiment, an attempt at Western-style public participation in distinctly non-participatory states still stuck in a Soviet mindset. But the gap between the good intentions and those flooded fields was starting to seem as wide as the 10,000 kilometres between Washington and Tashkent.

In Uzbekistan, despite the privatization rhetoric, the government was reluctant to loosen its tight grip on agriculture. Most Uzbek farms were still effectively *kolkhoz*, collective farms, operating more or less as they had in Soviet times. The government sent out decrees to farm managers who had to meet the quotas, still chiefly for cotton. Farmers were essentially labourers and had little input into the system. They followed orders and subsisted mostly on intensely cultivated vegetable plots and orchards. There were no incentives for farmers to conserve water. Water charges in Uzbekistan were flat-rate fees to cover transport costs, too small to matter. And water allotments were generous enough that farmers could operate as if water was endless. So they continued to flood their fields.

Within these restricting parameters we had to come up with innovative ways to encourage conservation. The teams' work so far was not likely to result in any real water saving. As Victor had said, the information was too much like Soviet-style propaganda. There were no call-to-action messages, no practical ways to cut consumption, and no rewards for doing so.

Despite Bozov's defensiveness about his teams' work, he seemed to understand this. One day he had raved to us about a water-saving incentive under Component A (water and salt management), a contest to see which water management district could save the most water. The winning district received as much as US$1,000, a huge amount of money in Uzbekistan. Bozov suggested we copy the idea.

"It means the water monitors have to be honest," said Victor sardonically when I asked him what he thought of the scheme. "Competitions are popular with authorities because they make them look generous. But they really just give them more power. And they use it to reward their friends."

Component B's indiscriminate budget was full of opportunities for cronyism. Each national team was allotted US$20,000 annually for journalists to write stories, broadcast authorities to produce programs and air messages and specialists to attend seminars and judge competitions. I already suspected that behind the team leaders' rants of "We don't need foreign specialists!" was the fear that some virtuous foreigner like me was going to check accounts and kill their reward schemes.

Poverty, greed and the huge underpaid bureaucracy lay at the heart of Central Asian corruption. Cronyism was just one type; kickbacks to bosses and government officials were another. Consultants and aid workers had warned me that in Central Asia bribes were routine. At the street level it was usual to pay off police for traffic violations, teachers and principals to get your kids into school, ticket agents to get a flight on short notice or a property agent to get a flat. But there was a higher level as well. Foreign operations in the region created a parallel dollar economy. Salaries for project staff – paid in American dollars – were ten times salaries for comparable skills in government or local business. And rents for foreigners' apartments were inflated by about the same amount. Giving a select few locals these windfalls created strong resentment among those who were kept out of the club. Foreigners working in the country were also expected to provide "gifts" or pay "fees" to their local bosses for landing contracts or doing business. According to development workers I'd talked to, it was the same story over much of the Third World, from Niger to Indonesia to Bolivia.

Victor saw corruption as immoral behaviour you must never stoop to. I was getting used to sitting in his car as he patiently endured the routine police checks in Tashkent. Their tactic was simply to delay you; a few dollars would allow you to be on your way. But Victor *always* refused to pay. And so we waited. "If you pay them you only spoil them," he'd say each time when he finally returned to the car.

But we couldn't take on this whole corrupt system. We had to introduce new methods that somehow avoided schemes that rewarded crooked officials and persuaded the real water managers to drop the sluices, patch the holes in the cisterns and limit the flooding of the fields.

"Soon you will see nothing but cotton plants," said Victor, pointing at another field. "You see, Robert, the problem with water in Central Asia is really cotton. As long as it's the major crop, water use is going to be excessive. If we really want to save the Aral Sea, we should be sending out messages to farmers telling them not to grow cotton."

The cotton shrub resembles a hollyhock and produces golf-ballsized fluff balls that can be woven into soft cool fabrics. It may look harmless but it has a long history wrapped up in colonial politics. Cotton has been grown since biblical times in Egypt, Arabia, India and the Americas, but it wasn't until Manchester entrepreneurs devised mass-production techniques in the early 19th century that demand for the downy fluff soared. Similar to other cash crops such as coffee, sugarcane and rubber, imperial powers grew cotton as a monoculture in regions where climate permitted and agricultural labour was cheap. They always exported the raw product home for processing. By 1860 the United States was the world's largest cotton producer, with the South completely dependent on slave labour to pick it. Even in those days the raw fibre ended up mostly in Manchester's cotton mills.

When the Confederate states lost the American Civil War in 1865, the world cotton industry collapsed. The resulting under-supply of cotton spurred tsarist Russia to expand cotton production in its new colonies in the Aral Sea basin, with the raw fibre sent to Russia for processing. A few decades later Soviet planners expanded the tsarist program, turning much of the Aral Sea basin into a cotton bowl, and again making little effort to establish a local manufacturing sector. Since the end of the Soviet era, foreign investors, leery of state controls and unconvertible currencies, have mostly stayed away and the five states have lacked the resources to set up cotton-processing plants. So in the year 2000 cotton was still Central Asia's biggest cash crop, providing a quarter of Uzbekistan's GDP, and still mainly exported as raw fibre. Cotton's colonial past was proving hard to shake off.

"We should tell them to grow wheat and other crops that don't use so much water," said Victor. "But the five presidents can't give up cotton. They're desperate for the cash it generates. They're addicted. They won't give it up until the land is ruined."

"Or the water runs out," I added.

"The Aral Sea Has to Die for the Red Army!" That wasn't a real Soviet slogan from the Cold War years of the 1950s, but it could have been. Because importing cotton was expensive, Moscow scaled up production in its Aral Sea "cotton bowl." Apparatchiks calculated the profits the crop would generate against the loss of the significant Aral Sea fishing industry. In Moscow the plan made sense, and through the 1960s and '70s it seemed to work. On top of their other rivalries, the Soviets and the Americans got caught up in a fierce cotton competition. In 1986 the USSR won: the Soviet Union became the world's biggest cotton producer. But it was straight downhill from there.

Because the Aral Sea basin lacks the necessary climatic conditions and water resources, many agronomists believe that cotton shouldn't be grown there at all. Cotton cultivation requires about 75 centimetres of annual precipitation; in a good year the Aral Sea basin can produce about 35 centimetres. Hence the huge demand for the waters of the Amu Darya and Syr Darya. In Uzbekistan, agriculture soaks up 90 percent of all available water resources. Cotton cultivation also requires extensive use of fertilizers and pesticides, which have now worked their way into the region's food chain, tainting everything from water and milk to fruit and vegetables. (This is not just a problem in Central Asia: while cotton crops occupy only 3 percent of the world's cultivated land, 25 percent of the world's pesticides are sprayed on the plants annually. Typically it takes 250 grams of pesticides to produce one cotton T-shirt.) The crop that had created America's shame, black slavery, was at it again more than a hundred years later on the other side of the globe. This time it was killing a sea and poisoning hundreds of thousands of people. The harmless-looking shrub knows how to bring out man's evil nature.

More than halfway to the city of Samarkand we traversed a narrow pass in some craggy yellow mountains called Timur's Gate. Tamerlane or Timur the Lame is Uzbekistan's tyrant hero. By the dawn of the 14th century he had carved out an empire that included most of today's Iran, Iraq, Syria and parts of Turkey, the Caucasus and northern India. In 1405 he charged through this gate on his way to conquer China. But he died shortly afterwards, saving many Chinese from annihilation and leaving the world his glittering masterpiece, Samarkand.

Tamerlane poured the spoils of war into his capital, turning Samarkand into a cosmopolitan centre of culture and spectacular architecture. (Later Napoleon, Stalin and Hitler would emulate his flair for autocratic city planning.) Many of his monuments are still evident today. He ordered canals dug, creating an extensive water supply system. He ordered artisans to create stunning cobalt and turquoise mosaics that embellished the massive portals, cupolas and minarets of the mosques and *medressas* (Islamic schools) where the city's youth studied, memorized and chanted the Qur'an. His citizens played backgammon under mulberry trees. They traded melons, apricots, pomegranates, grapes and apples in his bazaars, stocked with produce from his orchards and vineyards surrounding the city. By the standards of the time, life in Samarkand was very good.

The Silk Road survived from the 2nd to the 13th century. During the glory days, Samarkand was the halfway point, offering caravanserais – large inns with stables for animals – financial services and an East-West mix of cultures, peoples, religions and goods. The Silk Road was a network of intercontinental caravan tracks linking China and Europe. Coming from China a trader would traverse the Gobi and Taklamakan Deserts, scale the Tian Shan Mountains, descend into the Fergana Valley and trek across the steppe to the oases of Samarkand, Bukhara and Merv on the edge of the Karakum Desert. From there one branch headed north through the Kyzylkum Desert and Khiva into Russia and another swung south of the Caspian Sea through Persia to the Mediterranean. Most of the route was – and still is – bleak and treacherous terrain. The horsemen and camel drivers had to deal with marauders, war, heat waves, snowstorms and earthquakes. The only relief was to be found in the thriving centres of the Fergana Valley and in oasis towns like Samarkand.

Samarkand would have continued to prosper if Chinggis Khan hadn't sacked the city and slaughtered most of its 300,000 inhabitants in 1220. After that the city of mud dried up and blew away. Timerlane revived it 200 years later and it flourished until the 16th century, when the sea route around the Cape of Good Hope usurped the Silk Road. Remote and isolated again, Samarkand gradually evolved into a regional centre. Today it is Uzbekistan's second-largest city. But it has never quite recovered from its triumphant past. It belongs to a club of once-great trading cities that includes Venice, Liverpool, Bruges in Belgium and Savannah, Georgia – all cities stuck in time warps, living romantically off old mercantile glories.

My first encounter with Samarkand occurred in May 1998 in a far western province of Mongolia. I was in Bayan Olgii, which paradoxically means "rich place," giving a workshop for local NGOs on public awareness techniques. My work done, my Mongol driver and guide, Sony – he said he was named after a TV – took me off to see the local sights. It was a superb spring day and Sony was in high spirits as he charged his rusting Isuzu Trooper across the Mongol steppe and into a craggy ravine where goats, sheep and camels grazed on lush grass under wispy willows beside a chattering river. He pointed at the black boulders that herders had piled up on the sharp hills. They resembled the shadowy shapes of men.

"To scare away wolves," said Sony. He was serious. He claimed it worked.

"Where are we going?" I asked him.

"To see Samarkand."

"Isn't that two thousand kilometres southwest of here?"

His answer was only a friendly smile, a favourite Mongol response.

We came out of the ravine onto a harsh gravel desert surrounded by blue mountains that dissolved into smoky haze; the Siberian forests to the north of us were on fire. We stopped at an outpost called Bayan Nuur ("rich lake," although it was neither rich nor on the lake that lay beyond the town). Sony asked for Samarkand's house – Samarkand was a man! – and soon we were sitting at his table sipping salty milk tea and nibbling sweet biscuit crumbs that Samarkand called, with a chuckle, "Kazakh Snickers." Kazakh carpets lined the floors and walls. Bayan Olgii is a province of Mongolia with a Kazakh majority. Samarkand wore a purple Adidas tracksuit, a jaunty Kazakh skullcap and a cocky grin. He was a businessman. While Sony swapped Chinese-made sneakers for *koumiss* (fermented mares' milk), I wondered if Samarkand had taken his alias to reflect his business prowess, or just because he liked the sound of the famous name.

After tea, we filled the tank with *benzin* and headed across a desolate plain towards a snowy mountain range that featured the 4,193-metre Tsast Uul (Snowy Mountain). A strong cold wind suddenly blew out of nowhere; we'd been hot most of the day. We followed a steep track up a small mountain next to Bayan Nuur Lake while I thought, only Mongols would attempt to *drive* up a mountain. Sony's Isuzu choked on its latest fill of *benzin* but he diddled something and got it fired up again. We sputtered up to nearly 3,000 metres on the bald peak. White-capped Tsast Uul and her sisters spread out around us. Far below, Bayan Nuur was an azure puddle. I shaded my eyes and looked to the west.

"So where's Samarkand?"

Sony pointed beyond the mountains to the wide blue horizon.

"There! Can't you see the famous blue cupolas?"

Smiling in the gloom of the Hotel Samarkand's lobby, all sweetness and light, were Aziza and Ferouza. Aziza was wide-eyed, shy and sweet – her name meant honey. Victor had arranged for her to be my guide. Ferouza looked more stalwart. She was Victor's local partner in the NGO business that had brought us here. After introductions Victor and Ferouza departed and Aziza and I headed out to see the famous sights. But as we left the lobby, I dropped my camera and smashed the glass lens screen on the marble floor.

"It is bad luck!" gasped Aziza.

Minutes later, as we approached Tamerlane's mausoleum, I felt in my back pocket for my new Uzbek accreditation card to present at the entrance. But it had vanished. (Shakhlo will kill me, I thought.) Aziza and I retraced our steps, searching the lobby and our route across the pavement. Nothing. I was baffled. I wondered if Mr. G had his spies playing tricks on me, even in Samarkand.

"Now you must pay foreigner price to see all museums," said Aziza, upset. This meant $1 rather than 25 cents.

Trying to instill some fun into our so-far failed outing, I asked Aziza to smile in front of Tamerlane's tomb. But I couldn't capture Aziza's sad smile because my camera's light-meter battery was dead.

"Three times bad luck," said Aziza, devastated. "And only lunchtime."

It was a hint. Gradually we recovered over beet salad, chicken *shashlyk* (kebabs) and tea with lemon in an empty little café staffed by solemn waiters dressed in black and white.

Aziza stared earnestly at me and toyed with her food. Her round face was in pain.

"What's wrong?" I asked.

She smiled weakly and blushed. She said that she'd had a dream about me. "In it you are Negro. You have hair like black sheep. I don't know what to say to you."

"I'm sorry I turned out to be a mostly bald white man."

She laughed. "There are many Negroes in America." Maybe she thought I didn't know this. She laughed again. "But not you!"

The ice broken, she told me that her husband, Pazhman, had died of a heart attack a year before. He had been only 34. She was raising two children alone. She earned about US$15 a month as a teacher and clerk at the university. She had no prospects. "Soon I go to America. Or Canada." Her big eyes brightened.

On our way to the Registan we stopped at a photo shop, where surprisingly I found a battery that fit my old Nikon camera.

"Our luck has changed," I said.

I took pictures and in all of them Aziza was smiling.

The medieval Registan is one of the unofficial wonders of the world. It is a plaza surrounded by three monumental *medressas*, each smothered in elaborate cobalt and turquoise tilework. Bursting flowers, roaring winged feline-like creatures, geometric designs and Arabic script colour every inch of the arched portals, bulging cupolas and tapering minarets. I stood in the centre of the plaza and let the place engulf me. "Registan" means the "place of sand" as sand was scattered here to soak up the blood after executions. I imagined sword-wielding soldiers in silk turbans and blood-splattered tunics hacking off the heads of prostrate chained bodies. Hoisting the severed heads on the ends of pointed stakes, they shook them while the crowds roared.

The biggest carpet shop in the Registan was called Samarkand-Bukhara Ipak Gilami. The Afghan shopkeeper, attempted to educate me on silk and wool knots and weaves, natural vegetable versus modern chemical dyes, Bukhara designs, which were really Turkmen, as well as Afghan, Iranian and Azerbaijani styles. I flipped through dozens and dozens of carpets. All I learned was that each country had its own distinctive patterns. I told Aziza I wanted a blood-red Turkmen. She smiled and told me that she had a beautiful handmade Turkmen carpet hanging on her living-room wall. It had been presented to her late father when he retired from his job at a state tractor factory. She would let me have it for a much better price than any of these hawkers. When she assured me that it was bloodred, we had a deal.

In the noisy bazaar I bought a kilo of dried apricots and we nibbled them as we walked around the Bibi Khanym Mosque next door. Unesco was restoring the structure and a huge idle crane hovered above its 35-metre-high front portal. Completed just before Tamerlane's death in 1405, this was once the biggest mosque in the world. But its engineering techniques were extravagant and untested and the dome cracked before eventually collapsing in an earthquake in 1897.

As we walked around the ruins, Aziza told me the story behind the mosque. She said that Bibi Khanym was Tamerlane's Chinese wife – as a Muslim despot he of course had many wives. While he was off conquering, she decided to outdo her fellow wives and surprise her husband by building a mega-mosque. The architect she hired was a dashing Persian who had studied under the designers rebuilding the Doge's Palace in Venice. But he fell hard for her and refused to finish the structure without a kiss. Bibi, torn between her fearful devotion to her ruthless husband and her lust for this worldly architect, eventually succumbed to a kiss. When Tamerlane returned he spotted the lovebite on Bibi's neck and her beautiful head was soon skewered on the end of a stick next to the architect's at the Registan.

"It should be a Hollywood movie" I said, imagining Nicole Kidman as Bibi, Antonio Banderas as the architect and Russell Crowe as Tamerlane.

Aziza laughed. "It is only fable!"

Our last sight of the day was the Shahr-i-Zindah, the Tomb of the Living King. We entered a maze of alleys twisting through sun-baked brick catacombs, half-crumbling and half-rebuilt. Aziza wrapped her head in a scarf and walked deferentially with eyes bowed. I followed suit, except for the scarf, trailing after her through the cool and sometimes crowded passageways that ended in claustrophobic grottoes. I tried not to bump my head or show any disrespect to the lurking spirits. The atmosphere was conducive to contemplation and we lingered. Then we snaked out, pausing while a young mullah hypnotized us for a few minutes by chanting words from the Qur'an.

"*Sto gram!*" Victor ordered the waiter – 100 grams of vodka. "This is the best place in town," he announced. His day had gone well. We were eating dinner at the Café Crystal.

The waiter was Russian, a blond Apollo with a chiselled poker face. He poured our vodkas. "To Samarkand!" said Victor. We toasted Samarkand. Ferouza eyed the waiter up and down and teased him. But he was indifferent to her charms. He was hovering over the table like a statue.

Ferouza scowled at him. "*Russkii!*" she said, dismissing him. But still he didn't flinch.

"*Sto gram!*" ordered Victor. The stoic waiter nodded and walked away like a robot. We laughed.

We ate meaty salads, skewers of beef, liver and mutton and spongy french fries. We drank more vodka. We toasted Tamerlane and President Karimov, Bozov and Mr. G, Uzbekistan and Canada. The robot cleared the table. Victor stood up.

"I have an announcement. Aziza will become Robert's personal interpreter in Tashkent. He will give her a flat in Tashkent and take her back to Canada at the end of the year." He grinned at me. "To Aziza and Robert! *Sto gram!*"

The waiter nodded mechanically and fetched more vodka. Aziza smiled bashfully. I frowned at Victor.

After dinner we piled into a wreck of a taxi that took us out to Aziza's apartment building in the decrepit suburbs. The elevator was broken and we climbed to Aziza's flat on the ninth floor. We drank beers and gazed at the blood-red Turkmen carpet on the wall. It was beautiful.

Victor and Aziza argued over the price I should pay. After they struck a deal, we tumbled back down the urine-stinking stairwell into the night's chill and staggered into an Armenian bar nearby.

It was decorated with plastic flowers and lurid posters of pastoral scenes in the Caucasus. We ordered a round of Russian beers. Ferouza and Aziza sang along with the Russian pop songs playing on a fuzzy stereo. Three *mafiya* types across the room eyed them. Ferouza glared back at them and scolded them loudly. The taunting and chiding increased. Victor ordered more beers.

I took Aziza aside and told her I would buy her carpet. We left the bar and again mounted the dark stairs to her flat. I took the carpet off the wall and paid her. When we returned to the street Victor and Ferouza were standing next to the same wreck of a taxi that had brought us there. The driver had hung around all evening waiting for us. I thanked Aziza.

"Good luck with America," I offered.

But she turned shy and despondent, just as she had been when I'd first met her at the Samarkand Hotel. (Before the end of the year Victor would report that Aziza and her two children had become residents of New Jersey.)

The next morning, before returning to Tashkent, Victor drove us to Samarkand's Jewish Community Centre. A queue of about a dozen people had formed in the courtyard, waiting to see the director. A huge Israeli flag hung on the wall over his desk. Two teenage boys, playing video games on a computer, eyed us skeptically.

Victor explained his plan to set up ethnic-themed bed and breakfasts in Samarkand; this was the NGO business that had brought us here. He proposed a Jewish B&B. Why not at the Jewish Community Centre? If they would provide the facilities, he would arrange funding through embassies and foreign donors.

The director was polite but unsympathetic. He said that he was helping local Jews emigrate to Israel.

Stalin had begun the program of dispersing ethnic groups all over the empire, a forced migration that created cosmopolitan Soviet cities like Kiev, Tashkent, Baku and Almaty. It had also served to reduce dissent. It was comparable to the huge immigrations to North America, although Stalin's were not voluntary. In Central Asia there were Russians, Ukrainians, Armenians, Tartars from Crimea, Germans from the Volga River valley, Koreans from far-eastern Siberia, Mennonites and Jews. But the Jewish populations of Samarkand and Bukhara were unique; they had been there since the 12th century, the offspring of merchants and traders who had survived centuries of institutionalized discrimination. But now, with the five states run by their ethnic majorities, all these minorities felt marginalized and those that could were getting out.

"Very soon there'll be no Jews left in Samarkand," said the director. "So what's the point?"

6   Bring Your Own Weapon

"First time?" The accent was Australian, the woman's smile teasing. I admitted it was. "Doosh is a barrel of laughs. Heaps of fun. Heaps and heaps of stuff to do." She rolled her eyes. "Of course there's a curfew every night. But you can go out, if you don't mind pissed Russian soldiers using you for target practice."

The Slavic stewardess, the same smartly dressed woman who had collected our tickets at Khojand Airport, swayed up the narrow aisle of the Yak 40 prop aircraft. She was trying to keep up appearances. Tajikistan Airlines had seen better days. She served us sweet tea in thin plastic cups, a cellophane-wrapped candy and a smile.

Moira and I introduced ourselves.

"We just had a month off," she explained. She was sitting across the aisle from me. "Australia and Bali. Now it's back to rubbery chicken and *laghman* soup." She glanced at the young man beside her. "Isn't it, Matt?" He looked up meekly from his book and nodded hello.

Victor was dozing in the seat beside me. Out the window the jagged peaks of the Fan Mountains, a branch of the High Pamirs, glistened under a blanket of fresh snow.

It was a Wednesday morning in mid-February. Victor and I were beginning a new stage in our work that would continue for six weeks: meeting the public awareness teams on their home turf and planning our work together. This involved reviewing their activities, assessing their materials and coming up with some common goals for the year ahead. First stop: Dushanbe, Tajikistan. Sabirjan, our new driver, had just delivered us to Khojand, a small city about 150 kilometres south of Tashkent and just over the Uzbek border in Tajikistan. Poor relations between the Uzbeks and Tajiks had led to the cancellation of all flights between Tashkent and Dushanbe, the Tajik capital, which meant we'd had to drive to Khojand and fly from there.

"What brings you to Doosh?" asked Moira.

"Public awareness," I said, feeling challenged by her irony.

"That sounds like fun. Public awareness of what?"

"The Aral Sea."

"Hmmm," she said. "Isn't that a long way from here?"

I gave her a brief geography lesson on the Aral Sea basin that told her Tajikistan was part of it. "The whole region's headed for a major water crisis," I said, smiling. I didn't want to sound too alarmist, or too earnest.

"Oh boy! Another disaster!" Her eyes twinkled. "The civil war. Food shortages. Now a water crisis." She leered. "I'll let you in on a little secret – I think the Tajiks are too intent on killing each other to care much about the Aral Sea." She winked. "But Matt might disagree. He *loves* Tajikistan. Don't you, Matt?"

Matt felt obliged to tell me that he was the financial manager of a project run by Mercy Corps, an American aid organization. His looks and accent were midwestern. He said that he had been a Peace Corps volunteer for two years in the Fergana Valley. "I do the easy stuff – the books. I stay out of politics."

"From what I hear the books are where all the action is," I said.

Matt smiled defensively. "We oversee our own financial management. No local implementation, no corruption." He made it sound like a slogan.

"It's no fun at all," said Moira. "I think a little corruption keeps things interesting." Matt looked down at his book. Moira grinned across at me. "Matt can speak fluent Russian," she said with false pride. "He uses it to beg the soldiers not to shoot us."

At Dushanbe airport dozens of war orphans swarmed us, thrusting out little hands. Their dirty faces were exotic and frightening. I had no Tajik money but gave them some Uzbek *sum*, which created a small riot.

"Don't restart the civil war," said Moira with her ironic smile. "See you tonight at the Czech Beer Bar. Best entertainment in town. And the Tajik band is pretty good too. It's bring your own weapon."

Talbak Salimov rushed up and hugged us. He shooed the orphans away with a few curt Tajik words and led us across the empty square in front of the airport to his car. Ricky Martin was calling out to us over a tinny PA system, *"Uno! Dos! Tres! . . . Olé! Olé! Olé!"*

Talbak Salimov was the curly-haired leader of the Tajik team. Before leaving, Bozov had advised, "Robert! Talbak has the very best public awareness team! *You* will learn from *him*!" Nadir had added, "He is the *very* best public awareness specialist in Central Asia." It seemed that Valentina had been demoted, for the moment anyway.

"Welcome to the Paris of Central Asia," said Talbak as we drove down the wide Prospekt Rudaki. Manicured evergreens and gold and ochre buildings decked in oriental trim lined the boulevard. "The Champs Elysées!" he said with sarcastic pride.

"I expected Grozny," I admitted, pleasantly surprised. Tajikistan had been gripped in a five-year civil war that ended in 1997 with a tentative peace.

We passed ancient sputtering Lada, Moskovich and Volga taxis while shiny new UN High Commission for Refugees Land Cruisers sped by. Packs of machine-gun-toting troops in the olive-green uniforms of the Commonwealth of Independent States (CIS) congregated at every intersection. Talbak made friendly grimaces and pointed out the sights: the national Library, the post office, the opera house, the presidential palace.

Dushanbe was as Stalinist as Tashkent was Brezhnevian. But its ornate buildings were elegant compared to Tashkent's brutalism. Tashkent felt outsized – its spaces too large, its buildings too heavy, its crowds too thin. But Dushanbe was neatly colour-coded, consistent and intimate. It was pretty. The snowy Fan Mountains rose like a painted backdrop behind the city's pastel forms. The fronds on the evergreens fronting the presidential palace were lush and well-tended, lending a fairy-tale touch. Faces on the streets were handsome as well – dark Mediterranean features, thin faces, Roman noses and haunting eyes. According to Talbak, Tajik women, once favourite brides for Alexander the Great's soldiers, were still the region's most beautiful women. They floated along the sidewalks in long kaleidoscopic dresses and matching head scarves. The men were in dark suits or striped robes and turbans. Many had a southern stare that was seductive and challenging. Talbak had it, and my northern radar registered it as trouble.

Stalin rebuilt Dushanbe as the capital of new Tajikistan in 1929. Like so many towns (30,000, I read somewhere) he named it after himself – Stalinabad. But under Khrushchev it reverted to its original name, Dushanbe, meaning simply "Monday," the day of the town's weekly bazaar. This trite name suggests a lack of inspiration that still dogs the city. Despite its prettiness, it lacks character. Most of the population landed here from somewhere else, involuntarily. Many were Tajiks from Samarkand and Bukhara unhappy with the Uzbek administration that made them second-class citizens. More recently thousands of refugees from war-ravaged parts of the country drifted here, swelling the population to more than a million. But stuck in a mountain bowl with only one awkward way in and out, and rootless thanks to its artificial beginnings, it's a spiritless place, a provincial backwater. It may be nice to look at but, as Moira had warned, it's deadly boring.

Dushanbe's dubious status reflects Tajikistan's reputation as the region's black-sheep state. "Taj" means Persian speaker, and unlike the region's four Turkic-speaking ethnic groups, the Tajiks speak a form of Farsi and are culturally linked to Iran. They are the region's oldest ethnicity, descended from Indo-European Aryans, and are very proud of their long history, much of it coloured by struggles against repression. In 1929 Stalin, drawing freely over the map of Central Asia with his pencil, created the Tajik Soviet Socialist Republic. But his squiggled line placed Samarkand and Bukhara, ancient Tajik cities where the majority speak Tajik, in Uzbekistan; Tajik nationalists still talk of taking back these two cities. Stalin left Tajikistan with several tribes, the Tajik majority ruling over a mix of Uzbeks, Afghans and others. It's a fractious mess, almost ungovernable. After the collapse of Soviet rule, tribal disputes pushed Tajikistan, like its southern neighbour Afghanistan, into a long civil war.

Always the poorest state of the Soviet Union, Tajikistan is less a nation and more an ad hoc collection of Central Asian leftovers. Its mountains rival the Himalayas in height – at 7,495 metres Mount Communism is the highest peak in the former Soviet Union, five-sixths the height of Everest. But the Pamirs are too high and isolated for much human settlement. The tribes that eke out lives here are desperate and rebellious. Most of the country is more than 3,000 metres above sea level and only seven percent of the land is arable. With no energy resources other than water, the economy survives on World Bank loans and by not paying its debts. In 2000 it owed more than US$1 billion. Nearly a quarter of the state's farms have been privatized, giving the bazaars produce again, and with considerable mineral wealth the country has potential. But after almost a decade of war and political instability, it remains the most desperate of all the former Soviet Central Asian states, and one of the most destitute countries in the world.

A tank was parked in front of the Averso Hotel, where we were staying.

"A government official was assassinated here two days ago," Talbak boasted as soldiers waved us through the gate. "But don't worry, it's a very good hotel. Cable TV and long-distance telephone. And excellent security!" He leered. "You will like it!"

After lunch we went to Talbak's office, a ramshackle room in one of the Stalinist buildings that looked much better from the outside. On the wall, next to a map of Paris, was the same poster I'd seen in

Bozov's office, of a pretty young woman in traditional costume sitting placidly next to a mountain lake. The Central Asian Mona Lisa.

"You see terrible tragedy of Aral Sea," said Talbak, smirking as he joined me in studying it.

"Is that supposed to be the Aral Sea? It looks more like a lake in the High Pamirs."

Talbak shrugged and laughed. "It's the Aral Sea if you want it to be."

Members of his team arrived and a skinny boy made Nescafés. When Victor asked Talbak's driver to take us to see a building where we might hold our workshop, Talbak jumped on him. "He is not available for you!" he shouted. "He's busy!" Victor calmly accepted Talbak's rebuke.

"He doesn't understand us!" Talbak said to me. "He's not one of us. You should not work with such a man." He made little effort to keep this from Victor.

"Why? Because he's Korean? Or Ukrainian?"

Talbak squinted, closing one eye in a long wink. "Robert!"

"Talbak, I thought we were going to work together."

"Of course. With *you*." He was sarcastic. "But not with Victor Tsoy!" He sneered at me. "Maybe you are going to be like William."

It was a threat.

"Who's William?"

"You don't know about William?" He was still sneering.

He told me that about a year before, the World Bank had hired William to help set up the public awareness teams. He was an American who had worked for some time in Kazakhstan. The World Bank wanted some outside supervision in forming the teams and William was supposed to oversee the process. But apparently things didn't work out.

"He had blond hair," said Talbak. "He looked like California surfer. You ask Mr. Bozov what happened to him." He laughed. "If you keep Victor Tsoy, it will also happen to you."

The Victor conspiracy was growing. I still didn't completely understand it, but I knew it was coming from Bozov and Mr. G. And Talbak, who had shown promise at our first meeting in Tashkent, now was very much on Bozov's side.

Victor and I left Talbak's coffee party and went to see John Barbee, the country director of Counterpart Consortium, an NGO funded by the United States Agency for International Development (USAID) that trains fledgling NGOs in Central Asia. Before joining our team, Victor had worked at their Tashkent office, instructing NGOs on Western methods such as team play, grassroots organizing and lobbying. Each country office maintained a database of local organizations. While Victor collected names of Tajik farmers' associations that were working to improve irrigation on the plots of land they managed, I discussed Tajikistan with Barbee.

A strapping man in his fifties, Barbee was a rancher from Colorado. He was intrigued by my job and quizzed me on the goals of the public awareness component. Unlike most of the people I explained my work to, he was encouraging. He was the optimistic American. He told me about the Colorado River, how California was siphoning its water to supply Los Angeles and the fruit and vegetable farms in the Imperial Valley. "California is not paying the real cost for the Colorado's water. Those cheap California vegetables should really cost much more. Water is becoming a hot issue all over the world."

I asked him about the water situation in Tajikistan.

"On top of everything else, Tajikistan is suffering a severe drought. The water supply situation in Dushanbe is getting desperate." He explained that the World Bank was starting an infrastructure project to upgrade and expand the municipal water supply system. The inflow of war refugees had doubled the city's population and these migrants needed access to clean water. Then he talked about the civil war. It had displaced much of the population, with local Russians escaping north into Uzbekistan and Russia and some Tajiks fleeing south into Afghanistan. "No one knows how many died in the fighting – 60,000 is the official figure but 100,000 is more likely."

The civil war broke out in 1992 soon after Tajikistan's forced independence. Dushanbe's ruling Communist elite, dominated by Uzbeks from the northern province of Leninabad in the Fergana Valley (its capital is Khojand, from where we flew), could no longer keep the country's four tribes at bay. After rigged elections, members of the three long-suppressed tribes – Kulyabis in the south, Garmis from the east and Pamiris in the mountains – seized power. In August 1992 they set up their own coalition government. But clan tensions soon erupted and civil war broke out, encouraged by Islamic forces in Afghanistan. The Russians intervened, worried that if Tajikistan fell to Muslim militants, Uzbekistan would be next. By 1998 Russia forced a shaky peace and propped up the compromise president, Imamali Rakhmanov. Ever since, his brutal regime has been keeping Islamic warlords under control only thanks to 25,000 CIS troops.

"Political manoeuvring, skirmishes, assassinations – the country's full of rumours," said Barbee. "The other four states are afraid of a

'Tajikistan.' It's the threat of a Tajik-type war that keeps the five presidents in power and ruling with an iron fist. They're ruining our chance to build the democratic institutions these countries need."

Back in Talbak's office, Talbak dismissed Barbee and the Counterpart Consortium with a scowl and a wave of his hand that mimicked Bozov's rebuffs.

"We have our own NGOs!" he shouted angrily when Victor showed him a list of farmers' associations and asked if he worked with any of them. "You don't know anything about Tajikistan!"

The next afternoon Victor arranged the tables and chairs for group activities in a small meeting room in a government building. After introductory speeches, he asked the participants to stand one by one and speak for two minutes about their neighbour, a technique employed by Counterpart Consortium.

"That's not our way," Talbak scoffed beside me. "He doesn't understand us!"

"Where are your other team members?" I asked. There were only three out of about a dozen members.

Talbak screwed up his face facetiously. "They are busy."

Later Victor and I asked the participants to each write down their definition of public awareness. From these ideas we hoped to reach an agreement on one overall definition. It was "warnings," the participants suggested. It was "giving out information." It was "producing booklets and radio messages." It was "understanding that water was sacred." I asked them if public awareness activities should change the behaviour of heavy water users so that water was saved. There were some murmurs of approval.

Talbak jumped up. "This is only A-B-C!" he shouted. "We already know what public awareness is. Foreign experts may have their own definitions. But they don't understand Tajikistan!"

A few of the participants nodded. Eventually we got a definition that most participants agreed was about right. It included changing the behaviour of heavy water users to save water. Talbak mocked our achievement.

"Maybe you are happy with this. But my team members must all approve it. How can they do that when some of them aren't here?"

Robert Ferguson

Huge stainless-steel beer vats lined one wall of the Czech Beer Bar. About a hundred young Russian soldiers in fatigues filled the tables and Slavic barmaids, the only women in the room, served them mugs of beer. A live band was playing a slow song, a Soviet Second World War anthem – "*Ofiseri, ofiseri ...* " About a dozen soldiers had formed a circle and with their arms wrapped around each other's shoulders they drifted around the dance floor hollering the sentimental lyrics. After that, the band played a lively Tajik tune and they fluttered around the dance floor whooping.

There was no sign of Moira and Matt or any other expats. Talbak was glum. He eyed me suspiciously.

"What happened at your afternoon meeting?" he asked sarcastically.

After the workshop Victor and I had gone to see Tesha Avazov, the Tajik national coordinator of the Aral Sea project. He held the comparable position to pomaded Abdurakhim Jalolov in Uzbekistan – it was up to him to give his government's approval on all Aral Sea project work in the country. He was Talbak's Tajik boss. He looked like an old Soviet apparatchik with big tinted glasses and fierce scowl.

"What did Avazov say about me?" demanded Talbak.

I shrugged. "Nothing in particular," I lied. "The meeting was just protocol."

Talbak was afraid of Avazov, and he had good reason. Avazov had informed me that Talbak was not a good team leader. He'd asked me to help him to choose a new one. I'd said that I couldn't do that, it wasn't my job. But I said that I wanted to cooperate with him. His eyes had lit up and the scowl evaporated, and I realized that he'd taken my remark to mean I was with him and against Talbak. Again I found myself being forced to take sides.

"Hey, mister!" called a Russian soldier. No more than 18, he was peeking out behind a pillar and leering at me. "Come! Dance!"

The band started up and we joined his two friends, forming a huddle. We rocked and kicked up our legs. The boys' beery breath was in my face, laughing and bellowing the words to the popular Russian song. We danced faster and faster, spinning and swaying until two of the friends suddenly broke free. One squatted in the centre of the dance floor and howled. A space cleared around him. He pulled his pistol out of his belt and pointed it at the ceiling. "Bang, bang," he yelled, falling backwards onto the floor. He caught his balance and leaped back up with youthful agility. He aimed the gun at the three of us, his dance partners. We backed off. The band and the dancing stopped. The boy pointed his pistol around the room, yelling "Bang, bang! Bang, bang!" His blond hair flopped over his forehead. He had a lean face, high Slavic cheekbones and scared blue eyes.

I slid behind a pillar for a few seconds before sneaking back to our table. A waitress was watching the blond soldier. She was cool; she'd seen this before. She was waiting for him to calm down. After several more random aims and more "bang, bangs," the boy froze. He looked dizzy. The waitress ordered him to a chair and he collapsed into it. His buddies gathered around him, trying to get him to give up the gun. But the boy wouldn't release it. Now and then he taunted them, pointing it at each of them. But he was losing enthusiasm. Still gripping the gun he chugged down a mug of someone's beer, then hung his head. When he looked up he was crying. Someone grabbed the gun.

I looked at Talbak. He had paranoia in his eyes.

"He doesn't like me! He wants to change me!"

It took me a few seconds to realize he was yelling about Avazov.

"You must write a letter. You must tell him that I am a good team leader!"

"Talbak, I don't want to get mixed up in Tajik politics."

Then he turned furious. "You *must* write a letter! If you don't, you are like William! It will be terrible for you!"

I studied him for a few seconds, feeling some of his fear. I knew that if he lost his job he'd be destitute, like most of the population of Tajikistan. No more power, no more privileges.

"Send me a draft of what you want in Tashkent," I said, thinking I had to diffuse our growing factions. Despite my reservations, Talbak had potential. He understood public awareness and what we were supposed to be doing. If I could win him over we might get somewhere.

Minutes later he was smiling sarcastically at me again and I was regretting my moment of weakness.

"You know his gun was loaded. Yes!" My former dance partner had pulled up a chair beside me. He was leering at me again, a cherubic baby face. The band started up again. "Hey mister, let's go! Another dance!"

But two of his buddies walked over to our table, grabbed him by the scruff of the neck and dragged him back to their table.

7   From the Kazakh Steppe to the Heavenly Mountains

The young Uzbek border guard hugged his shoulder weapon and jabbed his finger at an open page in my passport. His eyes were fierce. His cheeks were mottled brown and yellow and covered in waves of peach fuzz. His green uniform, several sizes too large, hung off his lanky frame and his frying-pan hat looked huge.

Victor got out of the car and studied the problem. It turned out to be my Uzbek visa. On our return from Tajikistan, a border guard in Khojand who had no stamp had scrawled in the wrong date, a date that *predated* the start of my visa. The other guards gathered around to gawk at the tiny mistake and ogle me, this foreigner with a problem with his passport.

It was the balmy morning of February 24 and it felt like spring. We were stopped on the Uzbek side of the Kazakh frontier, less than an hour from home. Beyond the customs station the Kazakh steppe unfolded to the horizon, a vast ocean of parched grass. The Uzbeks turned their back on it, viewing it with contempt: too treeless, too untamed, too boring. They considered it a wasteland you couldn't cultivate, although the Soviets had worked hard to change all that.

Sabirjan, our driver, was taking us to Kyzyl-Orda, the next stop on our Aral Sea basin tour. Kyzyl-Orda, a small city 800 kilometres north of Tashkent and 200 kilometres southeast of the Aral Sea as the bird flies, was home to Dauletyar Bayalimov's Kazakh public awareness team. Unlike the other national teams, which were located in the safe confines of their capital cities, the Kazakhs had placed their team close to the action. Kyzyl-Orda province was in the Aral Sea Disaster Zone, an area delimited by donor agencies to draw world attention to the crisis. (In the back of my mind I'd hatched a plot to drive past Kyzyl-Orda and on to the shore of the infamous sea. There I'd take off my shoes and socks, dabble my feet in the brackish waters and christen this mission. But it was unlikely; there was no paved road and we didn't really have time.)

The holdup wasted most of the morning. The young guard knew that he'd stumbled onto something, but even his superior officers couldn't sort it out. Victor finally called the Uzbek Ministry of Foreign Affairs and convinced an official there that we would have the error corrected on our return to Tashkent. "No problem," he said, climbing back in the car, as if it hadn't been.

A couple of hours north of Tashkent, an island of smokestacks rose out of the sea of khaki steppe. Chimkent was a hodgepodge of industry: a lead smelter, an oil refinery, asbestos and cement works and a coal-burning power plant. Happily, a blustery wind was blowing off much of the smog. Over a bluff we descended into a fusion of old Soviet blocks, bright gas stations, flashy retail outlets, new deluxe hotels and restaurants and billboards shouting about cheerful products. Chimkent, capitalist and booming, was a landscape I could make sense of; it could almost have been a city on the Canadian Prairie where I'd grown up. With the scorched plains fanning out around its buildings and the snowy Talassky Alatau Range shimmering in the background, Chimkent looked like a Central Asian Calgary. It was wide open, vulgar and welcoming.

Lunch was mock Pizza Hut pizza and Chimkentski beer.

"Nothing like a little capitalism to liven things up," I said. The restaurant was busy and its buzz felt familiar, like in an eatery at home. Tashkent restaurants never felt like this. "Uzbekistan already feels like another planet, a thousand light years away."

Victor switched into his other personality, the one I'd met in Samarkand. He flirted with the waitress. He kidded Sabirjan, who was eyeing his pizza with curiosity, not sure how to eat it.

"Robert!" said Victor, impersonating Nadir's snide jeer. He wagged his finger at me. "You must *always* do what Mr. Bozov says! Mr. Bozov is your *friend*. He will *help* you! Together you will be *very* successful in Central Asia!"

After Soviet Uzbekistan, entrepreneurial Kazakhstan was liberating. Things were really changing here, maybe not all for the better, but its cheerful casualness was seductive. The world's ninth-largest country, Kazakhstan is the size of Western Europe, a huge chunk of steppe and desert with a scattering of lakes. It dissolves into Siberian taiga in the north, abuts the Caspian Sea on the west, skirts the Gobi Desert to the east and confronts a blockade of mountains on its southern edge, where we were now sitting. Its history reads like onslaughts of human tidal waves, each flowing over the steppe and washing away the previous empire: the Saka, the Huns, the Blue Turks, the Karakhanid Turks and lastly the tsunami of Chinggis Khan and his Mongolian Hordes. Out of these tribal flows emerged the Kazakhs in the 15th century.

Robert Ferguson

Kazakhstan is as inextricably linked to Russia, geographically and historically, as Canada is to the United States. Canada and Kazakhstan share predicaments; they both have vast barren territories and stores of resources that their powerful neighbours covet. The tsarist Russians grabbed the Kazakh steppe from the Three Great Hordes in the mid-19th century and the Soviets sent in new human tidal waves: Slav colonists to settle the "Virgin Lands," as its fertile steppe was dubbed, and dissidents to fill the new gulags they built in the wastelands. They also drove millions of nomadic herders onto collective farms where they died of starvation and disease by the hundreds of thousands. Those that fought collectivization were slaughtered. Between 1926 and 1933 Kazakhstan's population dropped by two million.

Empty and remote, Moscow chose the Kazakh Soviet Socialist Republic to test its most toxic experiments. During the Cold War the Soviets established the Baykonur Cosmodrome at Leninsk near the Aral Sea, the launching pad of the Russian space program even today, and tested nuclear weapons at Polygon, a stretch of steppe in the country's northeast. Today both Polygon and the expanding Aralkum wasteland approaching Leninsk are ecological disaster zones, full of sick and dying people, victims of radiation and toxic dust. And so far, politics and lack of money have prevented the cleanup from happening.

President Nursultan Nazarbaev, the country's former Communist leader, is now a capitalist dictator. He has pushed through "shocktherapy" economic reforms, selling off 17,000 state enterprises and opening up his country to foreign investors, notably oil companies. In 1998 he moved the capital from urbane Almaty to the windsweptcity of Akmola, renamed Astana ("capital city" in Kazakh) and once known as Tselinograd, the centre of Stalin's Virgin Lands expansion. Because Astana is geographically central, the move was seen as a ploy to placate the Slav majority in the north and ensure that the Russians keep their hands off Kazakh territory. Like the other four Central Asian presidents, Nazarbaev has smothered his opposition and kept a tight control over the media, which he also privatized. He now holds the title President for Life. The chief beneficiary of his economic reforms, he has become one of the richest men in the world.

At Chimkent we left the M-39, the highway that rolls on to Almaty, Kazakhstan's largest city, and turned north onto the M-32. A signannounced: "Kyzyl-Orda 530 km; Samara [Russia] 2012 km." The steppe stretched on to the cloudless sky. The only sign of topographical life was a hazy range of blue hills that crept along the eastern horizon.

A few Kazakhs, scrounging for fuel, chopped at the undernourished saplings that lined the highway, leaving only stumps and a few mangled branches poking out of the dusty ground.

"You see," deadpanned Victor, "the Kazakhs hate trees. The Soviets planted them so now it makes them very happy to cut them all down." He and Sabirjan swapped grins.

A policeman ahead waved his baton at us and Sabirjan pulled over. Victor and I stretched our legs while Sabirjan reported to the guardhouse. A white Volga with Russian licence plates pulled over beside us and four young Russian men got out and pissed into the breeze. One of them tossed an empty vodka bottle into a salty ditch. Another opened up a fresh bottle and offered the Kazakh officer a swig. He accepted and they shared a joke. Then the Russians zoomed off without a problem.

Sabirjan came back and said that the Kazakh police were demanding a letter spelling out who we were and where we were going. They had never heard of the Aral Sea project, the International Fund for Saving the Aral Sea or the World Bank. Our Uzbek diplomatic plates were the problem (Bozov had claimed the plates would guarantee us unhindered travel). We didn't have any letter.

Victor went into the guardhouse and half an hour later he got back in the car. He puckered his moustache and squinted with contempt.

"Kazakh police are the *worst* in Central Asia. Corrupt *and* stupid."

I was tired of the holdups. While I was waiting I'd noticed several cars drive by, ignoring the militia. Nothing had happened.

"Let's not stop any more," I suggested. "Tell Sabirjan just to drive past the next SIR checkpoint. (SIR is an acronym for something in Kazakh; it's pronounced like the Russian word for cheese, *seer*.) What are they going to do? Shoot at us?"

Fifty kilometres up the highway Sabirjan gunned the car past the next SIR post. He watched nervously in the rear-view mirror while Victor and I looked behind. The officer waved his arms at us several times and then gave up. We snickered at our impertinence.

"I think we've had enough Kazakh cheese for one day," said Victor.

In the morning I looked out the hotel window at the murky waters of the Syr Darya. The Syr Darya rose in the Naryn headwaters of the Tian Shan ("Heavenly") Mountains in Kyrgyzstan and wandered in a 3,000-kilometre path across eastern Uzbekistan and southwest Kazakhstan before trickling into the so-called Little Aral Sea. It was wide and shallow here, drifting sluggishly through ochre mud bars. Gangly willows struggled to grow on its banks. On the far shore the fields were plowed into banked squares, ready for flooding. In a couple of months they would be rice paddies. Kyzyl-Orda province, with the climate and terrain of Nebraska, was once known as the rice granary of the Soviet Union. It was still a major rice producer, mainly because the water here was too salty to grow anything else.

Dauletyar Bayalimov was at his office in the centre of town, sitting around the table with the members of his Kazakh public awareness team. His bloodshot eyes were circled in laugh lines and his elfish ears poked out. He told us about their work. He was in turns jovial and funny, exasperated and impatient, forceful and self-righteous, then boastful and rude. He would be enjoying himself and then suddenly become completely fed up.

The only woman in the room sat next to me. Gulira, young and willowy, was the administrative member of the team, the secretary, and also my interpreter. As Bayalimov spoke, she started to translate, but each effort trailed off. Now and then she smiled at his jokes. She seemed to forget about me and then to remember and start translating again. But she always gave up. After a while I told her that I wasn't understanding anything. She thought that was funny and smiled; all her teeth were gold, like Nadir's.

Lunch was *shashlyk* and vodka, which seemed to be the only food going in Kyzyl-Orda. Victor and I didn't want vodka but we had fallen under Bayalimov's spell. So we drank it. The vodka cheered up Bayalimov tremendously.

"Robert," he joshed through Victor, "have you ever seen the Aral Sea?"

Here was my opportunity. "No, not yet. Why don't we go and visit it?"

Bayalimov roared at my suggestion and then eyed me slyly.

"Last year Bassat wanted to see the Aral Sea, so I took him to see it."

José Manuel Bassat was a communications specialist from World Bank headquarters in Washington. He was monitoring the public awareness component and coming to Tashkent in a week.

"We got into his World Bank Land Cruiser and drove up the highway past the Baykonur Cosmodrome where Yuri Gagarin was shot into space in 1961." He smirked at me. "But probably you don't know about our heroes." He poured us more vodka. "We kept going to

Aralsk, Aralsk that used to be on the Aral Sea. It used to be our fishing port. But now it's nothing." A chortle. "We'd already driven 470 kilometres. Then we drove down onto the dry seabed, past a fishing boat stuck in the mud. We drove for another 30 kilometres, a huge cloud of poison dust trailing behind us. Then we stopped and got out. I said to Bassat, 'There's your Aral Sea!' It was flat, hardly any waves. No view. Nothing to see." Another roar. "Bassat stared at it. He walked along the shore for five minutes. I saw him put his hand in it. Then we got back in the Land Cruiser and drove all the way back. The trip took all day." Another chortle. "Now the World Bank knows the beauty of our Aral Sea!"

After lunch we toured the city. Kyzyl-Orda means "Red Capital" in Kazakh; it had acted as Kazakhstan's capital from 1925 to 1927, before the railway was extended to Almaty, which took over the job until Astana got the honours. A thousand kilometres to the east, across Kyzylkum Desert and the Big and Little Aral Seas, was the Tenghiz oilfield on the Caspian Sea. Because Kyzyl-Orda was the closest city, a few foreign oil companies had opened offices here. The Swedes had even built a new motor hotel, a sort of stripped-down Quality Inn; it looked as if they'd run out of money. (The rooms cost US$85 a night, but it was empty and the two front desk clerks looked lonely.) There didn't seem to be much of an oil boom going on. The city had no bustle. It was slow, windswept and desolate. The streets were full of dusty potholes.

"Water is being saved!" boasted Bayalimov back in his office.

We looked at his booklets of quotes from the Qur'an, his colourful calendar and dozens of newspaper articles. He claimed that the farms in Kyzyl-Orda province were now mostly privatized and that the farmers paid water fees. The combination of private farms and water charges was transferring responsibility for the land to private farmers.

Maybe. But the parched grassland around town, carved into ditches and furrows, looked more suitable for cattle than for cultivating rice. In the late 1980s large-scale rice paddies occupied 85 percent of the arable land in Kyzyl-Orda province and supplied 75 percent of the Soviet Union's rice. As it was with cotton, Moscow was striving for self-sufficiency. And like the cotton farms farther south, the farms here abandoned their less thirsty vegetable and wheat crops in favour of rice. Soon the usual over-irrigation problems arrived and salt began ruining the land.

In 2000 rice was still king in Kyzyl-Orda. This was because the water supply, the Syr Darya, was too salty from upstream discharge to nurture anything *but* rice. Rice thrives in salty water but demands even more water than cotton. The water statistics were frightening: A 100-hectare rice farm annually soaked up 2.8 billion litres, enough to fill 1,300 Olympic swimming pools. Producing the crop was only possible because water cost a piddling 65 cents a swimming pool. Kazakhstan still had 65,000 hectares of rice paddies, swallowing up an obscene quantity of precious water. Growing rice in the barren terrain around Kyzyl-Orda was an absurd proposition any way you looked at it.

"If you were as ugly as me, then you would be happy to sit in the dark," said Bayalimov. The power had failed and we were sitting in the dark in a private room of a popular restaurant. "I like this much better!"

We toasted the darkness – Bayalimov, two members of his team, Victor and I.

"But Daulet, think of all the pretty girls in the bar we can't see." It was Victor, in his happy Samarkand persona.

"It's okay for you," said Bayalimov. "You're handsome. They like to look at you. But they look at me and they want to run away!" He snickered.

The waitress brought candles, *laghman* soup, meaty salads, *shashlyk* and more vodka.

"Robert," said Bayalimov, his impish face flickering in the candlelight, "do you know how the Americans destroyed the Soviet Union?"

But before I could answer he told me.

"Public awareness! They used the Voice of America to brainwash our young people into believing in the wonderful American way of life. The funny thing was, it was all true. They weren't lying! That's the difference between Soviet propaganda and your public awareness – Westerners only tell the truth! Western people are very sneaky. They don't lie. We never figured that out. We thought lying was the way to go. Now we know, thanks to Westerners, that we were wrong. Tell them that the Aral Sea is dead! Tell them that we're wasting all our water. Tell them that soon there will be no water left for their children. Soon the cotton and rice crops will fail and we'll be hungry and broke. Tell them that. Tell them nothing but the truth. Thank Allah for the truth! Thank Allah for public awareness!"

We drank to public awareness.

The next morning our heads throbbed and Victor's stomach was off. He stocked up on *kefir*, a yogourt drink, which he swigged all the way down the M-32 to Chimkent. A mean wind blew down from Siberia. The sky was low and threatened snow. Our false spring was over.

At Chimkent we turned east onto the M-39. We hugged the foothills of the Talassky Alatau Mountains. Coming through a gulch, we hit a snow squall, then a whiteout. The highway was slick with black ice and we spun out, pirouetting and landing half in the ditch. We sat there dazed for a few minutes. Nobody said anything. Suddenly the weather cleared and Sabirjan drove all the way to Bishkek without stopping.

It was winter in Bishkek. Piles of snow lined the crumbling streets called Moskovskaya, Kievskaya, Sovietskaya, Frunze, Toktogul and Kyzyl Octyabri (Red October). It coated the black trees that grew willy-nilly through the sidewalks and smothered the Ukrainian-style cottages, the neighbourly *gastronoms* and the ramshackle blocks of low-rise flats. Bishkek looked good in snow; it covered up its shabbiness. Compared to Tashkent, it felt condensed and human. People plodded the pavements at a slower pace. The cars crept gingerly down its streets, skirting huge potholes. Bishkek didn't put on airs. Politics and ideology seemed to take a backseat to the everyday business of getting by. No one seemed to care that Lenin still waved from his soapbox in Ala-Too Square. The city was disorderly and agreeable; Westerners tended to like it. It was about as different from Tashkent as you could get while still being in ex-Soviet Central Asia.

Returning to Bishkek for me was like a homecoming. In March 1999 I'd lived here for a month developing a public awareness project for UNDP Kyrgyzstan, similar to the one I'd run for two years in Mongolia. In those four weeks I'd downhill skied and hiked through Ala Archa, a gorge only 30 kilometres from downtown, which ends at the Ak-Say Glacier. I'd also made new friends: Gabi Buettner of Zurich was planning new treks for us, this time on horseback into the mountains, and Bruno de Cordier of Ghent had just bought a new Niva 4X4 and had some hidden valleys for us to explore. I'd also left a suitcase of clothes with Bruno, my Mongolian winter gear that I suddenly needed more than ever. Tashkent's winter was nothing like this.

I directed Sabirjan to Valentina's office at the corner of Toktogul and Manas Streets. On the way I recognized old haunts – MacBurger, the Bombay, the Yusa Turkish restaurant, the Adriatico Paradise, the Bar Navigator and the American Pub. Bishkek's swelling expat and tourist populations were spawning Western-style eateries and bars. The city was popular with donor projects. In Central Asia, Kyrgyzstan was the easiest country to work in. And the world's second-highest mountains beckoned.

Sabirjan kept losing his way. He muttered curses about the snow, the decrepit roads, the police and the Kyrgyz. The Kyrgyz are too laid-back for the Uzbeks. They are Central Asia's mountain men. For centuries they have herded their sheep up and down the mountain valleys. In the summers they still pitch their yurts in the higher pastures. Their nomadic background has made them tough and independent. They enjoy life. They have round faces, warm smiles and unassuming ease, like their cousins the Mongols.

Valentina Kasymova introduced her team, three Kyrgyz men with big heads and big smiles and one skinny man who looked like a North American Indian. They were obliging. But Valentina was huffy. I told her that her office was the best we'd seen (I was being honest). She accepted my compliment with chilliness. But it was followed by a glint of appreciation.

Valentina was a serious and dedicated woman who followed orders; I'm sure that she would have followed mine if Bozov and Mr. G had told her to. She was thoroughly Soviet in a post-Soviet age that she refused to accept. Like many middle-aged officials in Central Asia, she preferred things just as they had been and carried on as if nothing had changed. Sometimes I though they hadn't. The transition to market economies in Central Asia was sluggish, even in liberal Kyrgyzstan. More than anywhere else in the former USSR, loyal Central Asian Communists had believed firmly in the Soviet brand of totalitarianism: the schemes and action plans, decrees and iron rules and privileges. Bozov's adherence to Soviet-style protocol continually amazed me. He lived with both fear and reverence for authority, and thrived on all the contrived methods necessary to get a decision made. I often wanted to ask him: "You do all that just to get approval?" "The Soviet Union was the best system in the world," he often told me with a wistful scowl. But at least he used the past tense. Valentina would agree wholeheartedly. Its demise was a shock they were still not over.

Valentina announced that she had been invited to attend a speech by President Askar Akaev. She would be busy all morning. I could feel Bozov's directive: *Don't cooperate with Ferguson!* She left Victor and me to spend the morning with her Kyrgyz team. They proudly showcased their work and lit up at any of our suggestions. Over tea they showed us a political cartoon in one of the daily papers that made fun of the prime minister – it showed him bumbling over some decision. Such audacity was unthinkable in the other four republics. Kyrgyzstan had the most open media in Central Asia.

It also had the region's smallest population and its second-smallest area. Kyrgyzstan is 94 percent mountain and has a mean elevation of 2,750 metres. With a snippet of irony, tour operators, embassy staff and aid workers like to refer to it as "the Switzerland of Central Asia." It is the most progressive state in the region. But only because independence in 1991 caught it in the lurch, too economically wobbly to sustain itself. Without resources such as oil or gas to trade, Kyrgyzstan was forced to turn to the International Monetary Fund (IMF), the World Bank and the United Nations Development Program. The government's liberal agenda has pleased donors, but left the standard of living for the masses sliding desperately. The four other brother states view Kyrgyzstan's progress snidely and with some jealousy, especially since Kyrgyzstan became the first country in the region to join the World Trade Organization.

The Kyrgyz team also showed us a newspaper article written by the deputy prime minister denouncing Kyrgyzstan's participation in the public awareness component of the Aral Sea project. Why should Kyrgyzstan support water-saving efforts?, he asked. With only seven percent of its land arable, water conservation was only an issue in the country's tiny enclave around Osh in the Fergana Valley. We should not have to save water, he declared. It was the Kyrgyz people's water and they were not wasting it. Uzbekistan and Turkmenistan were wasting water. Kyrgyzstan, he said, should be charging *them* for using *its* water.

Since independence the five Central Asian states had been bickering over the resources they'd once shared – oil, gas, coal, electricity and water. In Kyrgyzstan, the Toktogul hydroelectric power station could generate 10 billion kilowatt-hours of energy a year, enough to light the Toronto area. Yet each winter the Kyrgyz were forced to hoard water and leave Toktogul's mighty turbines idling, creating rolling blackouts across the country. This was because they were obliged to open their sluices in the summer months to provide water for the millions of hectares of cotton fields downstream. This agreement, signed in the early 1990s, basically continued the old Soviet arrangement. But these days the downstream states – Uzbekistan, Kazakhstan and Turkmenistan – were demanding hard currency for their gas and oil, leaving the broke Kyrgyz complaining and resentful.

They claimed they were losing more than US$90 million a year in lost power generation. Some Kyrgyz wanted to charge their neighboursfor their water. But Kazakhstan, Uzbekistan and Turkmenistan were balking at paying for water, the gift from Allah. With these kinds of grudges and jealousies abounding in the region, water was becoming a weapon. And without new fair-minded agreements, some analysts were predicting the region would devolve into water wars.

At noon Valentina returned and took us to the Café Classic around the corner for lunch. It was full of Kyrgyz, Slavs and expats, eating, chatting and laughing.

"It's much more relaxed here than in Tashkent," I said.

Valentina eyed me skeptically.

"You don't know what it's really like here," she chastised. "Things are falling apart in Kyrgyzstan. Democracy and capitalism are only helping a few people get rich. Everywhere there is crime. People are getting poorer every day. In Soviet times there was equality. Nobody was starving. Nobody had to sell their goods on the street!"

After the meal she ordered Kyrgyz brandy. One of her smiling team members winked at me. Valentina was softening. Later in the afternoon someone in the office produced a bottle of vodka. Valentina laughed and the bottle was opened.

The American Pub had taken over the foyer of the Drama Theatre on Chui Prospekt. The owner was an ex-Peace Corps volunteer from Texas and the menu was Tex-Mex. Teenage girls in tight jeans and "The Pub" T-shirts unloaded trays of draft beer, burritos, hamburgers and french fries. Newspapers and magazines – *The Times of Central Asia*, *The International Herald Tribune*, *Time*, *The Economist* – were spread over a table.

"Things are going better," said Victor. Over Siberskaya beer he was deconstructing our Kazakh and Kyrgyz tours. "We're starting to reach Bayalimov and Valentina. We're starting to build their trust. Soon they'll come around to our side."

Our burritos arrived and Victor poked at his suspiciously. Once he'd tasted it and found it agreeable, he continued.

"I understand Bayalimov now. I know why he drinks and why he makes so many jokes about himself. It's because his heart is broken. He told me that his son died a few years ago in an accident. After he told me that he admitted everything – about Bozov and Mr. G and their games. He said it's their games and he has nothing to do with it. He said they don't want BDPA. They want to stop the contract."

He signalled the waitress for more beers. The room was crowded, mostly with expats. It was smoky and noisy. The high ceilings, white mosaic-tile floor, oak-panelled columns and long mirrored bar gave the room sophistication. The old drama theatre foyer was a great space. It was built in the neo-classical style and the Soviets had not cut corners. And the Texan had done a good job making us feel that we could be in a bar in London or Sydney or Dallas.

"And Valentina?" I asked.

"You see," he said, eyeing a Peace Corps group at the next table, "to reach a Soviet woman like Valentina you must compliment her. You must tell her she is the very best. She is the very best public awareness expert in Central Asia. She is the very best water expert. She has the very best office. She has the very best team." There was a twinkle in his eye. "Then, when she knows she is the best, you give her flowers." He chugged the last of his beer as the waitress set down new ones. "What a Soviet woman wants above all is flowers. And roses are best. A dozen roses."

He glanced at me and puckered his moustache. He was being facetious, but he was also being serious.

"And Robert, when you are explaining your ideas to Valentina, you should remember – always sit to her left. It's very important. You see, she is completely deaf in that ear and she won't hear a word you're saying."

77

8   Mr. G's Baby-Blue Sea

Shakhlo was sitting cross-legged at the squat table on my front balcony, pretending to sip tea. The table looked like one you'd see in a traditional Japanese restaurant. She set down her make-believe teacup and smiled.

"You see, Robert?" She was demonstrating how to eat Uzbek style.

I saw but told her I couldn't sit like that to eat. "My legs are too long."

But she didn't care. She stretched luxuriantly and rolled onto her side, half onto the blood-red Turkmen carpet I'd bought from Aziza in Samarkand. She struck a pose that made her look as if she were a member of a sheik's harem.

Shakhlo and Faizillo, her driver – also her nephew, she said – were showing me my new flat. It was on the second floor of one of those five-storey blocks I could see out my office window. It was the day after I had returned from Bishkek and I was moving in.

"Please, you don't touch these things," Shakhlo said moments later, pointing at a desktop computer, a ghetto blaster and a heavy safe stored under the balcony floor. "These are private."

Faizillo was fiddling with the old Soviet air conditioner that was stuck through a balcony window. The view beyond him was of a quadrangle between apartment blocks: children's swings, an empty wading pool, a scattering of trees and some hibernating grapevines. Suddenly the air conditioner choked to life, rattling loudly.

"It work good!" Shakhlo smiled. We felt a blast of air, but it wasn't very cold. She turned serious and warned, "You can't see TV when air conditioner work. Because then TV don't work."

Soon Faizillo was in the living room channel surfing, through Russian and Uzbek stations, CNN and the BBC. He stopped at an English-language movie channel where Geena Davis and Brad Pitt were cavorting half-naked on a motel-room bed.

Shakhlo sat down on the sofa and crossed her legs. She eyed me quizzically.

"Robert! Cable TV! Now it's like home. Like Toronto."

Faizillo was standing with his back to her, blocking her view of the movie. She cocked her head and ran her eyes up and down his back. He was scrawny and looked more like a teenager than a man. But the strength of his sinewy body showed through his jeans and T-shirt.

Shakhlo said something that sounded teasing in Uzbek and smiled to herself. But Faizillo didn't respond. He was engrossed in Geena and Brad. Then she began kneading the back of his calf with her stockinged foot. Still he didn't react, but he didn't move away either, and she kept rubbing.

The next day my new interpreter started work.

Irina Vovchenko – Ira ("ERE-ah") – came to me through Tim Grout-Smith, a BBC media trainer I'd worked with in Mongolia. On my way to Tashkent in January I'd caught up with Tim in London. We'd reminisced about our media tour through the Gobi Desert as we helped raise public awareness of some endangered species such as the wild ass and the Gobi bear. An avid birder and amateur botanist, Tim had delighted in the summer flora and fauna of the Gobi, frequently ordering our driver to stop so that he could leap out of the Land Cruiser to photograph a clump of edelweiss or gaze up at a Houbara bustard soaring overhead.

"It's probably illegal to bring these in," Tim had said, passing me a paper bag of narcissus bulbs in the Chinese buffet in Golder's Green. "But please give them to Ira. I promised them to her."

He'd worked with her when he was shooting a video about the Aral Sea in 1996. The head of the project had insisted he interview farmers in the cotton fields.

"They were all well prepped, raving about their excellent crops and their first-rate irrigation systems. What a sham! Completely ridiculous. We gave up." He'd laughed a little bitterly. "But Ira's terrific. If she's available, I'd snap her up."

It turned out that Victor knew Ira and one evening he arranged for the three of us to go to the Ilkhom Theatre, Tashkent's experimental theatre. We saw a Swiss drama company interpret two plays by Euripides. The first was in German and the second in Russian, so I concentrated on the impressive staging.

At intermission we looked at dozens of black and white photographs of laughing, smiling and crying faces. The collection on display in the theatre lobby captured the reactions of Soviet citizens to the collapse of the Soviet Union. I asked Ira how she had felt. She said she was at Lake Issyk-Kul in Kyrgyzstan at the time, on holiday with her family.

"We weren't smiling or laughing. We weren't happy at all. We were shocked and upset." Her father worked for a Soviet company and they didn't know what would happen to his job. "We were scared. Nobody knew what the Uzbeks would do."

Ira had a delicate face, a slender nose and a small mouth. She was lean and poised. Occasionally she smiled warmly. But as she told me this, her expression was joyless and a little hard.

"My father still has his job, but things are much worse now than in Soviet times. We just want to leave. But we were born here. This is our country too."

After we'd looked at all the photos she said, "This exhibit only tells one side of the story. It's the side the West wants to see."

Ira was working with a Danish businessman who imported goods from Europe. The business was not going well. The Dane was trying to sue the Uzbek National Bank, claiming they'd stolen a million dollars from him.

"He's probably right about the money," she said unsympathetically. "It was some sort of exchange trick, but he's a crazy old man and I want to get away from him."

When I returned from Bishkek she called and said she could start right away.

"The Dane and I had a falling out," she explained. "Sooner or later everyone has a disagreement with him. I just hurried mine along a little."

"We don't want to listen to the foreign specialist! We don't want to listen to Victor Tsoy! We have heard these ideas before! We have our own ways of doing things in Central Asia!"

We were listening to the rants of a man named Ravshan Abdullayev. The Uzbek national coordinator, the pomaded Abdurakhim Jalolov who had advised me to reawaken Uzbek traditional respect for water, had sent his deputy to our workshop with the Uzbek national team. About forty, Abdullayev had a mop of unruly black hair and a vicious sneer. Over and over again he kept interrupting us. The members of the Uzbek team said nothing, but they continually nodded.

Victor, Ira and I faced them at the front of the room. We were in the National Law Library in central Tashkent. It had received some Canadian funding and in thanks the librarians had strung up some banners of red maple leaves over the bookshelves. Bozov was slouched in a chair at the side of the room, doodling on a scrap of paper and barely containing a smirk. Beside him was Nadir, gloating.

Sitting in the back of the room was José Manuel Bassat, the World Bank communications officer from Washington. He was here to monitor our progress. But he could see there wasn't any. Periodically he looked at me with dismay and compassion. This was not the way things should be going.

In his late twenties, José was short and dark and wore small oval wire-frame glasses that made him look stylish and intellectual. He was from a Jewish family from Barcelona; his father was in advertising. As a teenager he'd gone to an international school in Wales. He'd won scholarships and studied at Ivy League universities. He spoke Spanish, English, French and Russian. He'd lived for two years in Almaty, Kazakhstan. "I understand Central Asians," he told me with conviction. "I have an affinity for them." It was José who had gone off with Bayalimov to find the Aral Sea. "That was one of the worst days of my life," he said when I asked about the trip. "That man is crazy."

"Abdullayev is horrible!" said José over lunch. We were sitting with Victor in an outdoor café on "Shashlyk Street," the pedestrian mall outside the National Law Library. "He's really nasty."

"He shouldn't even be here," I said. "He's not a member of the Uzbek team."

We watched people stroll past and chewed on our kebabs. Shashlyk Street, also known as "the Broadway" – in Soviet times it was Karl Marx Street – was supposed to be fun. But it wasn't anything like Broadway in New York and it wasn't all that much fun. The cafés were indistinguishable. Each one had a red, yellow and white striped canopy hanging over a dozen white plastic tables and chairs. Out in front was a smoking *shashlyk* barbecue or a greasy cauldron of plov. The menu was four types of *shashlyk* – mutton, beef, chicken or liver – and one variety of mutton-topped *plov*.

Along the street buskers mimed, violinists fiddled, ballerinas pirouetted and hawkers offered tacky art, trinkets and souvenirs they'd laid out over the pavement. Passersby sat for their portraits for a few Uzbek sum. Kiosks sold bootleg CDs – the latest from Madonna and U2 were only US$2.50 each. Tashkent's youth, dressed in modest teen garb of Turkish Levi's, tracksuit jackets and Adidas, cruised up and down looking for something, anything. Girls, most in longish dresses, looked reticent. Older men handed out cards advertising opportunities with "beautiful ladies." But despite all this activity, Shashlyk Street didn't feel very festive. Maybe that was

81

because the Uzbek KGB headquarters filled a huge building at one end and the *militsiya* were everywhere.

"He was sent here to disrupt things," said Victor. He was in his sober analytical persona. "It's more sabotage. You see, this time Jalolov sent Abdullayev to wreck our workshop. It's interesting. It means Jalolov and Mr. G are *both* working against us."

Victor explained that Jalolov and Mr. G were old rivals. It seemed their mutual distaste for our contract was actually bringing them together.

"Well, it's good to know our work here is starting to have some positive results," I said.

"I don't understand," said José. "Why are they so threatened by your workshop?"

"It all seems to be orchestrated by Bozov, but he's under orders from Mr. G." I explained what had happened on the road, in Dushanbe, Kyzyl-Orda and Bishkek. "It's their campaign to undermine us." I smiled. "If they would conduct their public awareness campaigns as effectively as this one, there would be hope for the Aral Sea."

"Have you talked to Mr. Guiniyatullin about this?" asked José. He was genuinely outraged.

"José, I've been here six weeks and he's still refusing to see me."

José and I spent several days going over BDPA's original proposal and my assessments of the work of the national teams. We came up with an approach to their work that combined my ideas on pilot projects that had worked in Mongolia with some of the teams' activities that were already underway.

"Now let's go see Mr. G," José said.

Covering one wall of Mr. G's office was an enormous map of the Aral Sea basin. The Aral Sea was baby blue and vast and completely intact. The map was obviously Soviet and at least twenty years old. I wondered why he kept it there. During that first long meeting with him, I came up with seven reasons.

The first reason, I figured, was that it filled nearly one entire wall and since the rest of the office was stark and undecorated, it provided the only attraction. It gave the room its character, its conversation piece.

The second reason was that it was a wonderfully detailed map. Every topographical bump, every dirt track, every ghost town, every salt flat was on it. Maybe they were twenty years out of date, but they were there all the same, recorded history, waiting to be discovered over and over again. I wanted to walk up to it and follow the contour lines into the Aral Sea depression, trace the shore of the pre-disaster sea and interpret the Cyrillic names of the victimized towns: Nukus, Qongirat, Tahiatash, Hojeli, Moynaq, Kazaly, Aralsk. I could have spent hours staring at it – maybe just as Mr. G did. Maybe Mr. G and I had something in common. I imagined myself standing in front of his huge map for most of a day while he went about his business hosting other Western consultants and World Bank officials or lambasting his staff. Now and then someone would politely ask, "Why is that man studying your map?" "That's Mr. Ferguson," he would answer. "He's from Canada. He loves maps."

The third reason I thought he kept the map up there was because here wasn't a newer one to replace it. This was possibly true. The Soviets were great cartographers but now that they were gone and the Aral Sea basin was chopped up by new frontiers and squabbling states, it was unlikely that an updated version on the same massive scale had been produced. Or ever would be produced. Of course satellite photos tracking the demise of the sea existed. And so did many other maps showing how the great inland sea had shrunk into the brackish lakes now labelled the Big Aral and the Little Aral Seas. But they lacked the wonderful detail of this map.

Which brings me to the fourth reason: Mr. G liked the *irony* of having the map up there. He was making a statement. It was nostalgia for the great Soviet experiment in cotton production, an experiment he was part of. And it was a way to poke fun at the World Bank and Western consultants who were so obsessed with the disaster. He wanted everyone who met with him in his office, as I was finally doing a few days after our ill-fated workshop with the Uzbek team, to appreciate that Mr. Rim Guiniyatullin was one of the masterminds behind the Aral Sea's demise. He wanted his visitors to consider the scale of the disaster by absorbing every one of those 66,900 square kilometres of water (the Aral was once the world's fourth-largest inland water body, after the Caspian Sea, Lake Superior and Lake Victoria). From the chair where I was now sitting listening to Mr. G, its baby blueness was vast enough to swallow up most of my gaze, and my attention.

Reason five: Mr. G, despite his reputation, was exceedingly dull to listen to. This, I was discovering, was true. He knew it and kept the map up there to give his visitors something to think about while he rattled on.

Reason six: denial. The Aral Sea disaster never happened. Western environmentalists made the whole thing up. The map is up to date.

And finally, reason seven: he didn't care about maps or interior decoration. It was just something to cover a blank wall. His secretary had put it up.

I was propped between Anatoly Krutov and José Manuel Bassat. Both men were nearly a foot shorter than me and I felt that this should have amused Mr. G: *How many little World Bankers does it take to keep one big public awareness specialist from falling over in Central Asia? Two*! Ira was in more neutral territory near the end of the table. Mr. G was sitting directly opposite me and had his eyes fixed on me most of the time. My inclination was to glare and scowl back in return, more fiercely than he, like in a staring contest. But the consequences might be serious, so I let my eyes drift beyond him to the map.

For a while I compared mean old Mr. G with young military Mr. G. Young Mr. G's portrait hung on the wall. He had placed it very conspicuously – it was hanging directly over his head across the room from me, allowing for easy comparisons. The young Mr. G was almost unrecognizable; in fact, the more I looked at him the less sure I was that it was Mr. G at all. (Sabit Madaliev, who later became my right-hand man, came up with the theory that it was only a picture of someone Mr. G thought he would have *liked* to look like, like one of those haircut photos in a barbershop.) Whoever he was, he looked about forty. He was clean-cut with a lean face and hard features. Militarily handsome. Strong and unrelenting. The only trait I could be sure they had in common was the piercing glare.

Anatoly and José had spent the morning with Mr. G. After lunch they had come to my office and told me that everything was going to be all right. He'd listened to them. At five o'clock in the afternoon we would all meet. "Robert," José, tired and relieved, had said, "He's accepted your approach. Now you can begin your work." We'd smiled and laughed and for a few hours I allowed myself the luxury of believing that what Mr. G had told them was true.

But of course it wasn't. The head of all components of the Water and Environmental Management Project for the Aral Sea Basin and the director of the International Fund for Saving the Aral Sea was not about to allow me to work. The only positive step he was taking was to meet me at last, but that was only to stare me into submission.

Mr. G chain-smoked, lighting cigarette after cigarette. The smoke drifted above us and hung in the air. There was no ventilation in his office. He played with his Marlboro box, tapping on the table, setting it upright and then knocking it over. On the table was a large onyx ashtray that he squashed his butts into.

He must have said a lot of things that afternoon – he talked for close to two hours. But even with Ira's excellent interpretation I either lost the flow of his words or concentrated too long on that map (others have admitted to me that after a session with Mr. G, they too found it impossible to remember much of what he had said).

What I do recall is that he said he knew interpreters at the World Bank's headquarters at 1818 H Street in Washington only made $1,200 a month. Rent would take more than half their income, leaving them with little to live on. He implied that the World Bank was cheap. (This was likely meant for José, who'd told me that a beautiful Kazakh woman working as an interpreter in Washington had recently moved in with him. "I can't believe my luck!" he'd said.) Mr. G also claimed that he'd saved the Aral Sea project thousands of dollars by hiring local specialists instead of foreign consultants on several other subcontracts. Saving money and not hiring foreign specialists seemed to be his principal interests.

José looked mystified by all this. But Anatoly, apparently used to Mr. G's ramblings, retained his perky smile.

Then Mr. G mentioned Victor Dukhovny. Dukhovny had been the technical director of the Aral Sea project until the chair of the IFAS board rotated to a Turkmen and Dukhovny was shuffled into the post of director of the Interstate Coordination Water Commission of Central Asia. He and Mr. G were old colleagues, and old foes. They shared forty years of history, working together to expand the irrigation network and dry up the Aral Sea.

Just before I left for Kyzyl-Orda, Dukhovny had dropped into my office. He was a sharp old man, close to seventy. Speaking in English, he'd skipped the small talk and got right to the point: "What are you going to do about the national teams?" I'd smiled and said that so far they were not very effective. I wasn't sure what else to admit. He'd said, "They're a joke. They will accomplish nothing. Come and see me. I will advise you." But Mr. G was now exhaling Marlboro smoke and claiming that the national teams were excellent.

"Victor Dukhovny should know," he said, glaring at me. "He helped choose them."

When Mr. G finally finished, José and Anatoly took turns placating him. They assured him that BDPA was not here to undermine his public awareness teams or his authority in any way. They said that the BDPA foreign specialists would only assist him in any way that they could. Hadn't they agreed that morning that they were all working for the same goal? And hadn't they'd also agreed that there had been a misunderstanding? BDPA's specialists were not water specialists. They were communication specialists. Politely they reminded him that he'd signed a contract with BDPA. Hadn't he agreed to my new approach?

Mr. G said nothing. He stared at both of them and then looked away, bored.

Then I spoke. I explained my ideas as I had a month ago to Bozov and the four team leaders at our first meeting. I outlined my approach, suggesting that these methods could result in real water savings. I said that we would soon see some success.

Mr. G glared back at me and still said nothing. Despite our pleas it was no deal. Mr. G was reneging on his morning agreement. He was rejecting my approach. Maybe he believed he'd never made the deal. Maybe he'd misunderstood. Maybe he was thinking about another component – wetland restoration, say – and that I was proposing interpretation centres for nature lovers. But most likely he just wanted us to know that he called the shots.

He rose from his seat, put on his dark glasses, the kind popular with Soviet apparatchiks, and picked up his pack of Marlboros.

"Just remember this," he growled. He looked at all three of us in turn. "I helped create the Aral Sea catastrophe and only I know how to fix it."

He held out his hand. As I held it, I was again struck by his limp grip. I averted my eyes from his piercing glare and gazed one last time at his wonderful map.

"Visit me again," he said, not unkindly. He bared his tiny yellow teeth. "My door is always open."

"He wants something," I said to José. "That's what it's all about. He wants something from me. Or more likely, from BDPA."

We were sitting at a table in Gruziya, a Georgian restaurant on Pushkin Street. It was an open two-storey space that I guessed was meant to resemble a vineyard estate in the mountains near Tbilisi. The walls were overlaid in flat yellow stones and trimmed in dark wooden beams. There were large colour pictures of mountains and vineyards and people dancing in the costumes of the Caucasus. There were trellises with plastic grapevines. Most of the tables were arranged into long rows to accommodate large groups. It was International Women's Day, March 8, a holiday celebrated everywhere in the former Soviet republics, and the restaurant was full. Victor, José and I were treating our women, Ira and Shakhlo.

The waiter brought carrot and cabbage salads with walnuts; *khachapuri*, yogourt and cheese bread; *chakhokhbili*, chicken stewed in tomatoes, coriander, mint and peppers; and *khinkali*, hotly spiced meat dumplings. Georgian food was famous throughout the former Soviet Union and Gruziya was one of Tashkent's most popular nightspots.

José filled up my glass with blood-red Georgian wine.

"I still can't believe that performance. He broke our agreement. He lied!"

"He's playing games," I said, tasting the wine, which was thick and agreeable. "He has no intention of cooperating with us. I'm sure now he wants something."

"I've already talked it over with Anatoly. He's going to suspend all Component B funding until they start to work with you. Things will change soon. They'll have to."

But I didn't feel very reassured. Mr. G had just proven he could trump anything Anatoly and José came up with. The old monster was holding all the cards. José and Anatoly were his marionettes, and I was his consultant kewpie doll, something for him to bat around with until he got bored.

"I'm going to suggest to Jean-Charles Torrion we get out of this contract."

José looked alarmed. "Rob, give it some more time. You're just beginning your work here. There is so much to be done!"

I laughed half-heartedly. "I don't want to end up like William." William, the young consultant Talbak had told me about, had been hired by the World Bank to help Bozov set up the national teams.

José frowned. "William," he sighed. "That incident was unfortunate. Bozov and Mr. G played a lot of games with him. He never had a chance. When I arrived to work with him, Bozov told me William had already quit. I'm not really sure what happened."

"I have a pretty good idea."

José brightened. "Hey come on! William had no one. But you have BDPA, Anatoly and me!"

The band was playing a Georgian song. Shakhlo, Ira and Victor had joined a long line of dancers. Everyone was kicking up their legs and snaking around the room.

"Let's get out there," I said.

9   One Nation, One Man

Bozov's plump face leered at me across the rows of seats on board the Uzbekistan Airways flight. "Robert! Robert!" he shouted. He waved and beckoned me like an excited schoolboy with a frog in his pocket that he wanted to show me.

I made my way up the aisle and sat down next to him. He introduced me to the passengers around us as his *"Kanadski drook,"* his Canadian friend. Then he launched into one of his *Kanadski* anecdotes, with the clever Kyrgyz outsmarting the guileless Canadian (uh oh, I thought, they're starting to ring true). His audience, mostly middle-aged Uzbek *biznesmen*, was amused but, I suspected, less with the puzzling anecdote than with Bozov the clown.

It was a Sunday morning in mid-March and Bozov and I were on our way to Ashkhabad, Turkmenistan's capital, to set up the new Turkmen public awareness team. Because there were only two flights a week from Tashkent to Ashkhabad, Bozov had arranged a route through Urgench, the capital of Uzbekistan's Khorezm province. From there we would take a taxi across the Turkmen border to Tashauz, where we'd catch another flight on to Ashkhabad. The trip would take all day.

When I'd questioned our complicated travel itinerary, Bozov had said, "Robert! You will see Khiva!" Khiva was a well-preserved fortress near Urgench once notorious for its barbaric khans, slave traders and marauding herdsmen. It was one of the Silk Road's three star attractions – the others being Samarkand and Bukhara – and I was keen to see it.

The trip was Anatoly Krutov's idea. After the Uzbek workshop he and José had come up with a three-pronged strategy to get Component B and BDPA working together: First, build a relationship between Mr. G and me; there were plans for regular meetings, "To break the ice!" as Anatoly had put it (he loved English idioms). Second, threaten to suspend funds for Component B if Bozov and his team leaders refused to cooperate. And third, get Bozov and me to bond. "Robert! Make friends with him," Anatoly had chirped. "The problem is you don't *trust* each other. After a few days together in Ashkhabad you'll be old buddies!" But when Shakhlo had informed me that Victor's Turkmen visa application had been rejected – he was supposed to come with us – I suspected Bozov was up to his old tricks again.

At Urgench airport, Bozov hailed two taxis; one was for us and the other was for the two desktop computers and printer we were delivering for the new Turkmen team. He ordered the drivers to the border crossing into Turkmenistan.

From the back seat I looked out at deep irrigation ditches that sliced across sun-baked clay fields. Many of the fields were encrusted in salt and dotted with brackish pools. A few crumbling clay fortresses thrust above the flat fields like half-washed-away sandcastles on a muddy beach. Khorezm is an ancient land. It has been inhabited since the sixth century BCE and for centuries was a major Silk Road trading post. The area was once a huge wetland, a chunk of the Amu Darya delta sandwiched between two deserts: the Kyzylkum ("red sands") and Karakum ("black sands"). Its capital, Urgench, lies about 300 kilometres south of the Aral Sea. For the second time I was close to it, but a trip across the desolate expanse of the delta, which I'd gazed at out the airplane window eight weeks before, was again eluding me. "Not enough time!" Bozov had shouted cheerfully when I'd proposed it.

We crossed the Amu Darya on a floating pontoon bridge. The river oozed across its wide muddy bed, sluggish and thin. The water level was so low we probably could have forded it. The Amu Darya has changed course many times over the centuries and more recently dwindled so much that the irrigation channels barely provide enough water for the local rice and cotton crops. Khorezm is part of the Aral Sea Disaster Zone which includes Karakalpakstan, the severely blighted region in Uzbekistan on what's left of the sea, Kyzyl-Orda province in Kazakhstan and Tashauz province in Turkmenistan, where we were now headed.

"What about Khiva?" I asked Bozov at the Turkmen border. He laughed and pointed back down the road behind us. Khiva would have to have to wait for another trip. So much for the third wonder of the Silk Road.

(Three months later I would spend a weekend in Khiva. It would turn out to be a well-preserved museum, empty of people and things, a walled town of faithfully rebuilt shells of palaces, *medressas*, mosques and minarets that left me awed and slightly bored. Khiva was lifeless. Most of my memories of it soon faded, although one incident stood out: the obsession of a young Japanese man staying at our hotel to visit the Aral Sea. After spending ten years attempting to get an Uzbek visa, he was successful and overjoyed to be finally so close to his lifelong dream. But no one would take him the last leg. The taxi driver I found for him told me that beyond the devastated town of Moynaq, once on the sea, lay only the bleak, dusty Aralkum Desert; there was no road. But after extensive negotiations the driver at last took the Japanese man's $500 and off they'd gone. Whether they made it, I'll never know. But I'd desperately wanted to join them.)

In the middle of a roundabout in the windblown Turkmen town of Tashauz was a huge billboard of a middle-aged man in a navy-blue suit. He looked unexceptional: bushy eyebrows, chubby cheeks and an expression that made him look as if he were trying to hide a mouthful of food. Below him was the slogan *Halk, Watan, Turkmenbashi* ("One People, One Nation, One Father of All Turkmen"). This dumbfounded-looking man was Saparmurat Niyazov, president of Turkmenistan, popularly known as Turkmenbashi.

Formerly the Communist president of the Turkmen Soviet Socialist Republic, Niyazov was now prime minister, chairman of the Democratic Party of Turkmenistan and recipient of the Order of the Hero of the Turkmen People. His formal title was Saparmurat Turkmenbashi the Great. His baffled mug graced statues and billboards, hung over the front door of every public building and embellished banknotes and coins. Every man in the nation sported a Turkmenbashi watch, a gift from the great leader. Turkmenbashi was the new name for a major town on the Caspian, a meteor, a television soap opera, a TV channel and a proposed reservoir for irrigation discharge. In a slick marketing move, Niyazov had rebranded himself Turkmenbashi and his trademark squirrelly face and "One Nation–One Man" slogan were more pervasive than any advertising campaign by McDonald's, Coca-Cola or Toyota anywhere in the world. In Turkmenistan you couldn't escape Turkmenbashi – he was everywhere.

Sodium-gas streetlights flooded the night. Snowflakes danced in the yellow air. On the smooth black expressway from Ashkhabad airport Mercedes, BMWs and Volkswagens zoomed by and I wondered if we weren't actually on an autobahn heading into Munich. Ashkhabad, "the city of love" in Arabic, was surreal. Gigantic fountains spurted misty streams into the falling snow, like giant aerosols spraying through a buggy summer night. Fat bearded palm trees with snow-topped fronds lined long avenues. Gold domes and white palaces materialized out of the white-speckled darkness, basking in floodlight.

Usman Saparov, head of the new Turkmen team and, more significantly, the technical director of the Aral Sea project, named the sights with modest pride as we drove by (as Talbak had in Dushanbe):

Turkmenbashi's palace, the Majilis Parliament, the congress centre, the Sheraton Hotel Grand Turkmen, the casino. Ashkhabad was being completely rebuilt. Its new structures were a flamboyant mix of Islamic modern and Stalinist revivalism, with a little Robert Venturi Las Vegas thrown in. In the centre of town a 75-metre tower called the Arch of Neutrality (Turkmenistan has proclaimed a foreign policy of "positive neutrality") was topped with a 12-metre gilded Saparmurat Niyazov, the homely guy in his suit with his mouth full, rotating benignly over his capital. Usman explained that Niyazov turned so that he always faced the sun. But in the dark, what did he look at? His feet? I gazed up to see if the Father of all Turkmen was bowing his head like a sunflower at night, but alas the blizzard had swallowed him.

Ashkhabad was Niyazov's fantasy, a cross between Disney World and Abu Dhabi. European architects and planners and Turkish artisans were making it happen; the Turkmen people could only watch it emerge. It was financed with gas money; Turkmenistan is sitting on the fourth-largest gas reserves in the world. But it was *borrowed* gas money. The pipeline that would deliver Turkmen gas to Western consumers was not yet built, and Russia, Iran, China and Western oil companies were all arguing over its route. Would it go under the Caspian, through Russia, across Iran, or slice through the Tian Shan Mountains into China? The development of this gas line had turned into a modern version of the Great Game. Every few months a new plan emerged. *Forbes Magazine* called a 2001 proposal to run a pipeline from Turkmenistan to Pakistan via Afghanistan "the pipeline from hell, to hell, through hell." But all the squabbling did not stop Niyazov from borrowing on his people's future to remake his capital.

Usman drove on through the confetti night, across the city and up into the Berzengi strip, a collection of state-run hotels the Soviets had built twenty years before in anticipation of the gas rush. We passed a sheik's palace hotel, a spaceship hotel and a hotel that replicated a Turkmen carpet (well, sort of). The team of architects had no doubt been sent to Western theme parks for inspiration. We pulled into the Candebil Hotel. The lobby was a vision of late Soviet modernity, a Communist fantasy future *à la* 1984. It boasted a huge marble and glass atrium, open five floors up with overhanging balconies. Snowflakes bombarded the skylight far above.

Bozov led the way across the foyer's challenging landscape, down a few stairs, up a few stairs, around a non-functioning fountain. He leered at me as we creaked up the elevator, the same model that carried us up to the offices of the Aral Sea project. My room was a vast suite, empty but for a fuzzy TV, a broken dresser and a sagging single bed.

I was about to sleep when a knock at the door brought me face to face with Bozov again. He was naked except for his big boxer shorts. He held out a tube of lotion, grinned sheepishly and made ape-like charades indicating that he wanted his back rubbed. I obliged, kneading his blubbery shoulders while he moaned gratefully. He had a kink and more charades demanded I wrench it out of him. As I drilled into his shoulder blade with my thumb I wondered if this was bonding with Bozov. Was this what it took to get some cooperation? It was a peculiar sort of trusting intimacy, two men in their underwear standing in a rundown make-believe Western hotel on the edge of the City of Love. But the rub done, Bozov snickered and departed and I felt that instead of bonding, he'd just got another one over on me. He'd got himself another *Kanadski* anecdote.

Usman's office was jammed with prospective members of the Turkmen public awareness team. He ushered me into an adjoining room and gave me a stack of their CVs. I sat down with Gulistan, my interpreter, and we started to plow through them. Minutes later Bozov interrupted with "evaluation criteria."

"Approved by the World Bank!" he announced cheekily. It was a fill-in-the-number checklist to measure years of journalism experence, years of water management experience, et cetera, with a summary for the total years of all experience.

"Older is always better," I said to Gulistan with a smile.

She winced. "In Turkmenistan we have great respect for our old people."

Of course my choices all failed Bozov's numbers game.

The Bar Toronto across the street from Usman's office had a laminated tent card on the table announcing in Turkmen, Russian and English: "For your further enjoyment TORONTO happy hours! Every Monday from 11:00 till 23:00."

"What a good way to start the week," I said to Gulistan. As it was Monday lunch, I ordered a half-price beer from the waitress.

"Why is the restaurant called Toronto?" I asked her. Pretty and dazed, she shrugged.

"It sound modern. Like great city."

I told her I was from Toronto. She looked doubtful.

"Why?" she asked.

"I don't know," I said. "I just am."

But she seemed unconvinced, as if she didn't believe a city called Toronto really existed. She smiled and walked off, probably thinking I was only teasing her.

After lunch I walked with Gulistan through Azadi Square, the centre of town. Pine trees lined most of the streets; Gulistan said that the pine was Niyazov's favourite tree. The snow had stopped falling but the sky was leaden. Mushy puddles melted into streams that trickled into the gutters. Despite the chill, the fountains were all gushing. Ashkhabad's night glamour had turned flat in the grey light of day. Turkmenbashi's dream city had a ways to go. Surrounding the gold-domed palaces were vacant lots and decrepit structures slate for urban renewal.

"Foreigners suggest to save them," said Gulistan about the old buildings. "But the president wants them all demolished. He wants nothing left from Soviet times."

"I've never seen so many fountains."

"Because we live in desert, the water makes us feel good." But she admitted that the fountains could only use filtered water. "Our drinking water is not filtered and tastes like sand."

Ashkhabad's history is a story of water and earthquakes. For centuries it was a minor oasis outpost overshadowed by Nissa, 10 kilometres away and the capital of the ancient Parthian Empire. In the first century BCE an earthquake flattened the town, but later it re-emerged as a Silk Road trading post called Konjikala. In the 1880s the Russians, having just claimed the region by slaughtering 15,000 Turkmens at nearby Geok-Tepe, appreciated its strategic location and water supply and connected it to the world with the Trans-Caspian Railway. They rebuilt Ashkhabad, creating a cosmopolitan city with some impressive architecture. During the Great Game tourists found it an exotic destination – a European-style oasis city. But in 1948 another earthquake shook it to the ground, killing more than 100,000 people, two-thirds of the population. The city was rebuilt again, this time in a more dreary Soviet style. Then, in the late 1950s, the Karakum Canal arrived.

"We shall turn deserts into flowering orchards," proclaimed the propaganda of the time. The Karakum Canal was one of those mammoth Soviet projects designed as much to swell patriotic socialist pride – and show up the Americans – as it was to supply Turkmenistan with much-needed water. Begun in 1954, it was the key component of Moscow's plan to turn the Aral Sea basin into the Soviet cotton bowl.

Throughout the 1950s, the Karakum Canal thrust westwards across the scorching desert. Labour was cheap and plentiful so little heavy equipment was used. It opened up parched oasis outposts, watered new cotton fields and led the modernization of one of the Second World's most backward republics.

"The Aral Sea has to die like a soldier in the war of progress!" shouted a slogan designed to spur the canal diggers in the 1950s. This was to become the Aral Sea's epitaph.

By the 1980s the Karakum Canal was 1,100 kilometres in length, terminating in a reservoir just short of the Caspian Sea. It was the longest irrigation canal in the world. But according to UNDP reports, the canal was so poorly built and distribution so inefficient that only a third of its water actually reached the cotton fields. Most of the canal was unsealed and water seeped into the desert or escaped from its surface into the 50 degree-Celsius summer heat. The waterway also silted up. And as elsewhere in the basin, the flooding of fields raised groundwater levels, leeched salt to the surface and soon ruined arable land.

Joop de Schutter, a Dutch resource management consultant I met in Tashkent, suggested that if this canal were to be blocked off, the Aral Sea would return to its 1960s shoreline in just six months. Perhaps an overstatement, but the Karakum was still sucking up a quarter of the Amu Darya's water. And closing off the canal was the last thing on the minds of Turkmen management. Despite signing regional agreements stipulating less waste, more prudent water management and quotas on sharing the Amu Darya with Uzbekistan, President Niyazov had given his go-ahead for the implementation of a plan that was part of the old Soviet irrigation scheme for the region. Construction had already begun downstream from the Karakum on the Amu Darya on a new canal, the Trans-Turkmen Canal. Eventually it would extend across the midsection of the nation.

Ostensibly this parallel Karakum was going to provide muchneeded drainage on irrigated lands now watered by the Karakum. But it would also open up a new fertile band for development, increasing the country's irrigated lands from 1.2 million to 5 million hectares and boosting its flagging cotton harvest, which in recent years had slumped to 75 percent of its peak in the mid-1980s. The new canal was to end at Lake Turkmenbashi, a huge desert reservoir of discharged water in the middle of the Karakum Desert, 1,000 kilometres south of the water-starved Aral Sea. The Turkmen planners claimed that not a drop of this reservoir's water would come from the Amu Darya. But many scientists and environmentalists disagreed. They were warning that this scheme would only extend land degradation by transferring polluted water from one area to another. They also said that it *would* bleed more of the Amu Darya, further straining Turkmenistan's already shaky relations with Uzbekistan.

Usman had some miraculous news about the nation's pre-eminent waterway: "Turkmenbashi has just announced that the Karakum Canal is no longer a canal. Because it has changed course and itsbanks have been altered, he says it's turned into a river!"

Wonders never cease in the magical realm of Turkmenbashi.

Just before leaving Ashkhabad, I was invited to sit before the chairman of the International Fund for Saving the Aral Sea, the umbrella organization that oversaw the Aral Sea project. Tekebai Altyev was also Turkmen national coordinator for the project. Since taking control of IFAS, he had become Mr. G's chief rival. Rumours were floating around that he was about to wrest control of the project away from Mr. G. Which meant, as things were going, that he was a potential ally.

Altyev was another old apparatchik. Like the other national coordinators – scowling Avazov in Dushanbe and pomaded Jalolov in Tashkent – he had perfected the surly style of the autocrat. He was stern and overbearing. I was there only to listen to him and not to ask questions. First he rationalized the necessity for the new TransTurkmen Canal. Despite his regional role as chairman of IFAS, he had no qualms about blatantly supporting this Turkmen scheme, even if Uzbekistan was outraged. Then he informed me that he was going to move the headquarters of the Aral Sea project from Tashkent to Ashkhabad. In line with this, I was to present all my foreign specialists to him for approval and hold all my training workshops in Ashkhabad. This was outside the terms of our contract, but of course I wasn't allowed to say anything.

Back in Tashkent a few days later, Bozov handed me a list of the members of the new Turkmen team. My top three picks had been a young woman from an NGO, a young man working in the government's printing house and a 40-year-old male reporter. None of them made the list. I had known they wouldn't; I had ignored the "evaluation criteria." The successful candidates were all males between the ages of 58 and 74.

Robert Ferguson

When I asked Anatoly Krutov about the World Bank's "evaluation criteria," he told me he'd never heard of them. When I asked him if it was necessary for me to parade my consultants before Altyev for approval and hold our training workshops in Ashkhabad, he just laughed.

"Of course not Robert! Mr. Guiniyatullin would never agree to that!"

10   45 Tazedinova Street

Sabirjan swung the Turkish Fiat north off Abdulla Kodiry Street and onto Usmon Yusupov Street. We passed a precinct of shops that filled the ground floor of an old Stalinist building that had somehow escaped the 1966 earthquake. There was a new *gastronom* that boasted Danone yogurt, French wine and a few European cheeses. And a few doors down a mini-Eiffel Tower was propped over the door of a patisserie. Inside were bread and pastries. No baguettes, but there was a gooey Uzbek chocolate eclair.

"*Malyenky Pareezh*," I said. "Little Paris."

"Robert, you're dreaming," said Shakhlo. "Tashkent is not Paris. It is *much* more beautiful." She laughed at her joke.

"You're right," I said. "It's Dushanbe that is the Paris of Central Asia, with its wide avenues, its beautiful palaces and its many, many gendarmes."

"Robert, please, give me your mobile. I must call about next house."

It was early April and we were looking at houses in Uzbek neighbourhoods, *mahallahs*, in a part of Tashkent known by its Soviet name, Rabochi Gorodok (Workers' Town). With four of our consultants arriving in mid-May, I'd asked Shakhlo to narrow down a list of possibilities to four or five houses I could look at that day.

"He will meet us there," she said, finishing her call but not returning my phone. "It's best house, Robert. You will like it. It has pool and sauna and fruit trees just like you want. Jean-Charles Torrion will be very pleased. You will see."

I hoped so. This would be the fourth house we'd looked at that morning. The others had not been right. The first was a ten-room palace with a huge pool and garage and a rent of US$2000 a month. The second was small and rundown. The third we could only peer at through the gate; its garden had turned to weeds.

Sabirjan turned off Usmon Yusupov into a narrow lane called Tazetdinova Street. A small statue honouring an Uzbek poet who had grown up in this *mahallah* marked the entry to the lane. Cherry, quince, apple and apricot trees were coming into bloom on both sides. New grapevine shoots wrapped themselves around iron trellises, their baby leaves opening. "*Asalam aleykum! Asalam aleykum!*" ("may peace be with you" in Uzbek) shouted some children playing in the street. They smiled and waved as they moved aside to let us pass. Several of the houses were under renovation, with new purple and pink marble facades and ornate hand-carved hardwood trims.

"This work Uzbek tradition," said Shakhlo, pointing at an intricately carved wooden door. "*Very* beautiful."

Near the end of the lane Sabirjan pulled over in front of the steel gate of number 45. It was a one-storey renovated house with outside walls stuccoed in tiny black, maroon and yellow pebbles. Wrought-iron grilles covered the tinted windows.

Another car drove up and a middle-aged Uzbek man got out and greeted us. He unlocked the gate with some difficulty, as if he hadn't been here before, shook my hand and directed me inside. "Please, welcome mister," he said.

The centre of the courtyard was a garden of oriental shrubs, young fruit trees and a basil hedge. A patio table and chairs were set out on the cobblestones under an extended roof. Opposite the main house, on the far side of the garden, was a guesthouse with a bedroom, bathroom and sauna. Beside it was a small swimming pool.

"This looks perfect," I said, smiling at Shakhlo.

"Of course, Robert. I told you."

We entered the house through the kitchen. With its wide stainless-steel gas range, high ceilings and wooden table for prepping food, it looked Italian. It opened onto a dining room with a glossy hardwood floor set out in the design of a four-pointed star, like the symbol for NATO. There were two bedrooms off the dining room and two living rooms down a hall. The first had a large TV; the second was enormous, empty except for a four-metre-long wooden table and a dozen chairs. The tinted windows at the far end looked out over Tazetdinova Street. The house was spacious and elegant, yet homey.

"This is it," I said to Shakhlo in the huge living room. "This room will be great for our workshops and meetings. And the other room with the TV is perfect for Phil and his crew." Phil Malone, our video production consultant, had already been in Tashkent training journalists for a European Union project, and he'd told me the facilities he needed.

Shakhlo eyed me nonchalantly. "I will ask price."

She went to talk to the property agent, who had been talking on his mobile phone most of the time. He treated her with slight contempt, as if he didn't approve of a woman being involved in *biznes*. In *Uzbekistan biznes* was a male domain. But he didn't intimidate Shakhlo. She challenged his condescending sneer and made her demands.

A grinning middle-aged man and a skinny youth arrived. They turned out to be Akhror, the owner of the house and coach of the national junior soccer team, and his son Alisher. Akhror shook my hand enthusiastically as if we already had a deal. Alisher spoke some English.

"If you have problem, please call my mobile," he said. We exchanged phone numbers. He looked at my card and smiled. "I am at your service, Mr. Robert."

"How much is the rent?" I asked Shakhlo in the car as we headed back to the office.

"$1,200," she said. Rents for foreigners were always in U.S. dollars in Uzbekistan.

"Did you try to get it down?"

She gave me a cool glance. "How much you pay?" Her tone was abrupt, as if I was bargaining with her.

"I don't know. What do you think?" But she had no opinion. "How about $900?" I suggested.

"I will try, Robert. But this man is difficult. Maybe he won't agree."

When we reached the office, I had to ask her for my mobile phone back.

A few days later, Victor, Ira and I were sitting with Talbak on the patio of the Czech Beer Bar in Dushanbe enjoying a lunch of *shashlyk* and beers under bright spring sunshine. We'd just completed workshops in Almaty and Bishkek with the Kazakh and Kyrgyz teams and Tajikistan was our last stop on this second round of training. The patio was crowded, not with CIS soldiers like the bar inside at night, but with laid-back government workers. Birds were twittering in the leafy trees around us and the balmy weather was keeping everyone from returning to work. And we had another excuse: we were celebrating Ira's 29th birthday.

"Ooooh, Robert!" Talbak snickered as he slid a chunk of greasy mutton off a skewer with his fingers. "Mr. Bozov is *very* angry with you." He deftly rolled the meat in the vinegar and raw onion marinade and popped it in his mouth.

"Well, I'm very angry with Mr. Bozov," I said, chewing on rubbery chicken.

Bozov had tried to ruin our workshops in both Almaty and Bishkek. He'd shown up at the end of our Kazakh workshop and shouted, "The foreign specialist is only here to steal your ideas!" In Bishkek he'd arrived late one afternoon just in time to hear Valentina announce: "Foreign specialists have no new ideas. We don't need them!" The next day a headline in one of the state-run newspapers declared: "Foreign Specialist Talks, But What's The Point?" The article claimed that when foreign specialists come to Kyrgyzstan they are only trying to steal the local specialists' good ideas.

The plan had been for us to rendezvous with Bozov in Dushanbe for the Tajik workshop, but I'd called Talbak and made up an excuse that delayed our arrival by a week. I was not going to hold another workshop with Bozov in the same city.

"He was waiting for you here three days. You tricked him!" Talbak was chastising but also delighted.

"Only after he tricked me by trying to sabotage my workshops in Tashkent, Almaty and Bishkek." I smirked at him. "But I know I won't have any problems here."

Talbak scrunched up his face and winked. "I *always* cooperate with you!" Then he turned angry. "Why didn't you write a letter to Tesha Avazov? You promised!"

He was right. I had agreed to send a letter to the Tajik national coordinator, who was threatening to replace Talbak as team leader. But with Talbak so entrenched on Bozov's side, and with Bozov playing tricks, I wasn't about to do Talbak any favours.

"You never sent me a draft." I smiled and swigged my beer.

"You don't know how to write a letter?" he sneered. "I can make a draft right now!" He pulled a pen out of his jacket pocket and began scrawling on a napkin.

"Talbak, if we have a good workshop here, I will write a letter to Avazov."

Victor was watching us with disapproval. Ira was ignoring us, looking dreamily around the patio, determined to enjoy her birthday no matter what.

"To Ira," I said, holding up my mug of beer. "Now she must decide which country in Central Asia has the sexiest men!"

"Of course Tajikistan," said Talbak, scrunching up his face again.

At the Tajik workshop no one said that the foreign specialist had no good ideas. No one said that I was going to steal their ideas. Now and then Talbak frowned sardonically at me, but he remained uncharacteristically quiet.

After the workshop Victor, Ira and I went to see Tesha Avazov. The Tajik national coordinator introduced us to an academic-looking man of about fifty, a respected Tajik writer.

"The new leader of the Tajik public awareness team," Avazov announced happily. Then he showed us a letter signed by the prime minister of Tajikistan that spelled out the new team; Talbak was still a member, but no longer the leader.

The new leader looked less than enthusiastic about the honour bestowed on him. Like me, he likely felt he was being drawn into a political dispute that he wanted no part of. His apprehension should have signalled trouble ahead, but blinded by Avazov's cooperation we charged ahead, coming up with a new work plan for the new team. When it was finished Avazov signed it and made it official. Then he hugged each of us.

"I can't believe it," I said to Victor that evening as we toasted our achievement at the Czech Beer Bar. "We might actually be getting somewhere. It feels strange, like it's too good to be true."

"I knew things would change for us," said Victor in his analyticalpersona. "You see, Avazov knew that Talbak was corrupt and he had to demote him. Now he's in control of the team and he knows that by working with us they can achieve some results. This will make him look good. We will start with the Tajik team. And when the other teams see our success they will cooperate. Soon, Robert, you will realize your dreams."

I wasn't so sure, but I was happy to accept his hypothesis that evening.

We watched Ira fluttering to the lively Tajik beat with a throng of gyrating soldiers in khaki uniforms. I'd never seen her so happy.

"I thought you were my friend!" Talbak was suddenly standing over our table. He had a hurt pout on his face, his grey suit was wrinkled and his tie was askew. "But no! You are not my friend. Now you and *Tesha Avazov* are friends. *Best* friends!" He sneered before adding, "You are a traitor!"

"Talbak, I'm very sorry for what has happened," I said, pulling out a chair for him. He ignored it. "I had nothing to do with it. Tesha Avazov changed your position."

101

Victor tried to explain the prime minister's letter, but Talbak turned on him.

"You! You don't understand us. You are not one of us." Then back to me, "Mr. Bozov will be very angry with you!" He threw back his head defiantly.

The music stopped. I watched Ira's smile wilt as she looked over at us and spotted Talbak. She wisely stayed put, waiting for the next song.

"Wait until you get back to Tashkent. Then you will see what you've done!" Talbak backed away from us and moved slowly towards the door. "Mr. Guiniyatullin will stop you! You will see *very* soon!"

"*Na zdorovya!*" shouted a Russian soldier, holding up his beer. The men around his table raised their mugs to Talbak. There were a few cheers and the band started up as Talbak disappeared out the door.

"Congratulations, Robert!" bellowed Bozov back in Tashkent. "You have restarted the war in Tajikistan! Talbak is the first casualty!" He pointed at Victor. "And *he* is giving you terrible advice. Get him out of here!"

Victor stared stubbornly back at him and refused to move.

"And *she* is not a good interpreter," Bozov yelled, sneering at Ira.

Nadir dropped his mouth and feigned shock. But Ira's icy stare cowed them both.

Bozov turned back to me. "You fired Talbak, the best public awareness specialist in Central Asia! But Mr. Guiniyatullin won't allow it."

Following Bozov's tirade there was an eerie silence. Bozov lit a cigarette. He blew the smoke out in my face and leaned back in his squeaky chair. His eyes were gleaming.

"Robert! I have a surprise for you!"

"A *really* big surprise!" added leering Nadir.

Bozov's really big surprise was an all–team leaders' inquisition. I was on trial.

On a sunny and warm afternoon in mid-April I found myself sitting in the project meeting room beneath the gaze of the five Central Asian presidents. Opposite me were my accusers, four of the national team leaders: Bakhtiyar Nazarov, his Uzbek chest puffed out; Valentina Kasymova, her round Kyrgyz face testy; elfish Dauletyar Bayalimov of Kyzyl-Orda, who looked drunk; and my new best enemy, scowling

102

Talbak Salimov of Dushanbe. Usman Saparov from Ashkhabad sat on the sidelines, looking as if he didn't want to be there. To my left sat the World Bank's tiny and gleeful Anatoly Krutov – my lawyer I deduced – and beside him Bozov the judge. Ira was my reluctant personal counsel. Bozov had ordered me not to bring Victor, the co-conspirator.

For nearly two hours the evidence was presented: I had stolen their ideas. I had failed to appreciate the excellent work of the national teams. I had conspired against one of the best public awareness experts in Central Asia. I had meddled in politics. I had refused to take the excellent advice of the director of Component B. I was a traitor. I was a double-crosser. I had no ideas of my own.

Anatoly winked at me. "They need to *let off steam*," he chirped. "Just let them *get it all off their chests*." I wasn't too sure about his defence strategy, but I thought if he was enjoying himself, then what the hell, maybe I should too.

Then a special treat: Valentina, in a demonstration designed to show Anatoly that they didn't need any foreign specialists, presented her Kyrgyz public awareness strategy. "This strategy," she announced, "can be used as the model for the whole region." Talbak held up a big chart made up of intricate boxes and circles joined by dozens of crisscrossing lines. Valentina explained her objectives and methodologies, target groups and actions for half an hour. Ira didn't interpret any of it. At the end, she whispered, "Robert, it's just impossible to explain."

When it was over we all applauded. Then Bayalimov announced, "Valentina is not only the best public awareness specialist in Central Asia, but the most beautiful woman as well!"

Lots of cheers and the door flew opened and Nadir arrived with a bouquet of a dozen roses, which he presented to Valentina. She stood up, embarrassed and delighted, and blinked away a few tears.

"*Spasiba bolshoi* [thank you very much]" she sputtered, smiling tightly. Another round of applause and she sat down and smelled her flowers.

In the end I was disappointed. Despite the severity of my crimes, I was not found guilty. Bozov the judge reaffirmed Talbak as leader of the Tajik team and assured everyone that only Mr. Guiniyatullin could change the members of a national team. (For several weeks after this I received faxes from Tesha Avazov in Dushanbe lobbying for my support in his battle with Mr. G over Talbak, but I tried to stay out of it.) But Bozov pronounced no sentence on me. No public beheading in

Mustaqillik Maydoni, the former Lenin Square in the centre of Tashkent. No extended stay at the new gulag President Karimov had just opened near the remains of the Aral Sea. No cancelled visas and one-way ticket to Toronto. No, I had to stay and continue my mission. They weren't going to let me off that easily.

Two days later Jean-Charles Torrion arrived from Paris. He'd brought Bordeaux for Bozov and Cognac for Mr. G. "We'll see if these little bribes help," he said brightly.

JC's slight build, boyish face and mop of wavy dark hair made him look younger than his forty-five years. He'd completed a master's degree at the Michigan State University in Lansing ("Such a boring place you cannot imagine!" he'd said, laughing) and later spent several blissful years working in Australia and Costa Rica. At BDPA he was director of development, second in command, working mainly in former French colonies in Africa. He had an easygoing nature and a slightly amused view of the world. He was most intrigued by our crazy project.

Despite his laid-back attitude, JC was very sharp on figures. He told me that he had memorized all the numbers in Russian. "When it comes to negotiating, this is most important. This is what matters." And it was our numbers that he looked into first.

"Do you know what she's doing?" he asked me, as he reviewed the accounts that Shakhlo managed. He was sitting at one of our computers in my office. "It looks like she's adding $50 to her salary each month. Incrementally. It's not that much, but it's starting to add up." He laughed. "She's stealing."

I called Shakhlo into the office. She sat down and scrutinized us as JC explained the discrepancy. Then she looked indifferent.

"I do only what Victor Tsoy tell me to do," she said coolly, looking at JC. "He say I must add this money every month. I was very surprised, but he tell me it is rule of Aral Sea project. He show me in contract that my real salary is $700. Right now Robert pay me only $500. So I just add $50 every month. Of course I stop at $700."

JC lit a Gauloise. He was trying hard not to smile.

"Victor told you to do this?" I asked. "Do you want me to ask him?"

She shrugged. "Maybe he forget."

JC explained to her that the amount in the contract was only a budgeted amount. "This is the maximum, Shakhlo. No one is paid the full amount. Not Victor Tsoy, Robert or even me." He smiled at her. But she did not smile back.

"I only do what my boss tell me. I think Victor Tsoy is my boss."

I looked at JC and chuckled. "Victor will be amused to hear that." She shrugged again. Then I told her that her salary would stay at $500 unless I approved an increase. She gave me a cold look, got up and left.

"You know, on many projects," said JC, "it's usual to fire the office manager after six months. It's the best way to stop the pilfering." He laughed and shook his head. "Such impudence! She is really unbelievable!"

But to me she had just crossed the line. Now I really couldn't trust her. I should have fired her right then and there. But I didn't. I liked Shakhlo and I still believed I could win her over to my side. But of course I was dead wrong.

Bozov fondled his bottle of Bordeaux and then plunked it on his desk. He grasped JC's hand and massaged it. "My very good French friend!" he said and tousled JC's hair.

Nadir took the bottle off Bozov's desk and stowed it under his.

"Doctor says Mr. Bozov must not to drink," he confided loudly to Ira and me. He put his hand over his rounded belly and made an ugly face. "Very terrible *sto-mak*!"

JC lit a Gauloise and laughed half-heartedly at Bozov's anecdote about a Kyrgyz, a Russian and a Frenchman, which none of us got. He put up with Bozov's pawing. Then he asked Bozov why he wouldn't approve my schedule for the workshops with our consultants in May.

Bozov looked surprised. He turned to Ira. "Maybe her translation is not clear." But Ira stared him down. He ordered Nadir to show him the Russian schedule. Nadir muddled through papers on his desk until he found it. Bozov looked at it as if he'd never seen it before. He grinned.

"It is possible! I must refer this to Mr. Guiniyatullin for his approval!"

Down the hall Mr. G accepted his bottle of Cognac graciously. On the matter of the schedule of consultants, he looked indifferent.

"This is a matter for Mr. Bozov," he said mildly. "He is director of the public awareness component, not me."

At the end of the meeting Mr. G rose from his chair and glowered at JC.

"I told you at the start of your contract that Victor Tsoy is not a water specialist. He should not work on your team. But you do not listen to me."

Then he stuck out his limp hand and said almost pleasantly, "I wish you luck with your seminars."

Outside his office JC and I grinned at each other. Mr. G had just given us his approval. The May workshops were on.

At 45 Tazetdinova Street the cherry blossoms were in full bloom. The apricots were coming along nicely and Alisher, the owner's son, said the quince would be out in another week or so. He was busy preparing the garden, meticulously raking up winter debris and planting a few rose bushes. The grapevine had filled out and now provided a broad canopy over one end of the patio.

Shakhlo was in the kitchen, taking stock of cutlery, glasses and dishes and compiling a list of everything we needed for our consultants. The list was growing.

She hadn't got us a deal on the house. She claimed there were others who wanted it. I doubted most of what she said these days, but JC had already approved it, so I took her word on the rent.

"I know cook and cleaner," she said. "Very nice sisters. They will be very good. You will see."

I asked her to show me receipts for everything she bought, with each item written clearly in English. In Tashkent few cash registers issued proper receipts and usually she provided just a scrap of paper with a note scrawled in Russian. Some were impossible to decipher.

"Remember, we have to send them to Paris and JC will look at them." I knew this wasn't much of a threat. Shakhlo fixed her eyes on me for a few seconds.

"Robert, please you will check every one very careful that I am not cheating." Her tone was slightly mocking. Then she looked a little hurt. "I try very hard to make everything perfect for our foreign specialists."

With preparations underway, I left for a few days' break in Bukhara.

11 Their Country, Their Way of Doing Things

Old Bukhara, the *shakhristan*, was a maze of cobblestone alleys lined with sun-bleached yellow-brick walls. There were no vehicles. My Zurich friend Gabi led us through a gate in a wall and into a sunlit courtyard surrounded by shaded alcoves where a smiling young man offered us tea. Gabi had a knack for finding things, hideaways like this, complimentary tea and handsome young men.

We ducked under a low doorway into the young man's cool nook and viewed his orderly display of printed fabrics, paintings and etchings. He was very polite. He said that next door were shops selling jewellery and ceramic figurines. And next to them were Tajik and Uzbek hats and long quilted coats and silk dresses. And after that "Bukhara" carpets. Gabi told him in Russian that she wanted only art and silk: one or two paintings, if they struck her right, and a bolt of the psychedelic silk that was sold by the metre. A seamstress she knew in Bishkek would sew it into a suit. "I'd only wear it on special occasions," she told me with a wry smile. We both knew that she might wait forever for one of those special occasions, but she wanted it anyway.

Gabi Buettner was 29. She was on a break from her duties as a junior program officer at UNDP Kyrgyzstan. She was from Zurich, but as a teenager had lived for two years in California. Her English was North American colloquial. She'd studied international affairs and Russian and after graduating lived in St. Petersburg to learn the language properly. When I first met her in Bishkek in March 1999, her co-worker had told me she spoke Russian like a native. "I'm quite good at languages," Gabi had admitted with a small smile. "I have an ear." She was fluent in German, English, French, Spanish and Russian. But her work bored her. "I'm a bureaucrat," she said. "I don't actually do much. Two years of this will be plenty."

After tea we followed a narrow canal that gushed beside one of the alleys. Several ancient hand-manoeuvred slots regulated its flow. The canal led to a square pool surrounded by ancient mulberry trees that were dropping their white and purple berries all over the paving stones.

The fat berries were full of seeds and tasted sour, a bit like raspberries. They squished under our feet and stained our sandals. In the shade of the mulberry trees, white-bearded men in gold and blue quilted coats and embroidered caps were playing backgammon. A few tourists and locals sat at tables next to the pool sipping tea or beer and idly chatting. In Bukhara the Labi-Hauz (Tajik for "around the pool") was where everyone ended up. As in Samarkand, the locals here were mostly of Tajik origin.

It was only early May, but already the Bukhara sun broiled. Each of our three mornings in the city we walked the alleyways before the midday heat depleted us. We crisscrossed the canals on elevated stone bridges. We inspected *medressas* and mosques. We climbed the Kalon Minaret, built in 1127. It was 47 metres tall and criminals were once put in canvas bags and thrown to their deaths from its top. I looked straight down, imagining them plummeting and splattering on the stone pavement in the square below. The canvas bags made for easy cleanups.

Bukhara had been one of the original Silk Road trading towns. By the 10th century it had become *Bukhoro-i-sharif*, noble Bukhara. Its Persian poets, philosophers and scientists had made it Central Asia's Islamic heart. It was well known in the Islamic world and today is still revered as a Muslim holy city.

Bukhara survived Chinggis Khan's routing in 1220 and later being overshadowed by Samarkand, Tamerlane's capital. Much of the *shakhristan* dates from the 16th century when it was the capital of the Bukhara Khanate and boasted a bustling bazaar and over a hundred mosques. The khanate was ruled by a series of cruel despots whose notoriety spread around the world. In 1842 one of them, Nasrullah Khan "the Butcher," beheaded Colonel Charles Stoddart and Captain Arthur Conolly, two British officers Queen Victoria had sent to Central Asia on a mission of peace. It marked the beginning of the Great Game. The Bukhara Khanate surrendered to the Russians in 1868. The Trans-Caspian Railway opened up the city to tsarist Russians in 1888, although the emir, Abdallahad Khan, was still powerful enough to insist that the station be built 15 kilometres outside the city. Bukhara still manages to keep a distance from the modern world. It felt pleasantly backward, not really part of President Islam Karimov's Uzbekistan. And I couldn't quite imagine it ever being part of the Soviet Union either.

From the top of the Kalon Minaret, Bukhara spread out around us. Baked a golden brown, its partly rebuilt buildings were crumbling at a leisurely pace. It was less showy than Samarkand (there were fewer blue cupolas) and much quieter; the vehicles were all relegated to Soviet Bukhara, outside the *shakhristan*. We heard only the echoes of children's shouts as they played in the streets and the calls of a few vendors pushing pomegranates and watermelons. A garden of red poppies bloomed in front of the Abdul Aziz Khan Medressa. A mishmash of mini-domed roofs, like a layer of bubble wrap, sheltered the warrens that made up the spice bazaar, the cap-makers' bazaar, the jewellers' bazaar and the money-changers' bazaar. On the edge of the *shakhristan* rose a massive bulge, a thick mud wall. The Ark, Bukhara's citadel, dated from biblical times. It looked like a gigantic flaxen whale. Late one afternoon we approached it as the sun was setting and watched it glow. The colours were brilliant – ochre sand, gold and pink. They turned deeper and then faded away.

Bukhara was agreeably sluggish, serene and timeless, and it made us feel that way. Its history was raw, staring us in the face. Its yellow-brick walls seemed to stretch time into a continuum, layers of naked records that laid bare everything that had ever happened here. I was surrounded by truth. I felt relief, as if I'd finally escaped from the sea of deceit I'd been drowning in. Bukhara was a refuge; it was releasing me from the clutches of Bozov and Mr. G.

At six o'clock on our third afternoon Gabi and I were at the Labi-Hauz drinking beer and snacking on *samsas*, an Uzbek version of a samosa, when the pool suddenly erupted. We were sitting with a group of Europeans who were all on breaks from jobs in Central Asia. Water gushed out of spouts along the sides of the pool. It shot in streams and splashed onto the pavement beside us. Everyone at the Labi-Hauz let out a collective whoop. In Bukhara flowing water, spurting water – such activity! – was astonishing. And wonderful. Someone technical dampened the thrill by explaining: they were only back-washing the pool, rinsing away the daily sediment.

The pool was still erupting when my mobile phone chirped. It was Victor.

"Robert!" he chuckled. "Bozov says, 'Robert did not come to Central Asia to have a holiday! Tell him that *again* I am waiting for him!'"

109

"Where are the foreign specialists' handouts? Where are their methodologies? Where are their CVs in Russian?" Bozov was on a rampage. Now that he had agreed to "let us work," his anxieties about our upcoming workshops overwhelmed him. He was rude and abrupt. He ordered me to formulate detailed action plans outlining every minute of each foreign specialist's visit. He presented me with his "Scheme for the Seminars" with 12 headings, including "Tools to be used to reach the target of the seminar," "Inciting factor for the seminar" and "Monitoring of the seminar's efficiency."

"I think he's serious about these workshops," I said to Victor. "Maybe he's not going to sabotage them."

"Don't count on it."

Ira was spending most of her time translating Bozov's almost indecipherable memos, which arrived daily. She gave me one that was five pages in Russian but only one page in translated English.

"I see that you've done some reducing," I said.

"You don't need to know most of what he said," she replied coolly. "Nobody does."

"Maybe Anatoly Krutov would be interested."

She shrugged. "Robert, you're taking your work too personally." Since her birthday in Dushanbe, her attitude had changed. She was now disgusted with Talbak, Bozov and Nadir and wanted no more involvement in our work than necessary. "You should detach yourself from all this," she said, looking at the memo. "Otherwise it will destroy you."

Shakhlo read a few of Bozov's memos. One day she offered me some practical advice: "Mr. Guiniyatullin is waiting. He say he want money."

On a Sunday morning in late May, I had breakfast in the cafeteria of Bishkek's Pinara Hotel with two World Bank monitors from Washington. José Bassat had brought his boss, Paul Mitchell, the chief of regional operations for external affairs. In an earlier e-mail José had said that, "unlike myself, Mitchell is a big tall Canadian – like you," suggesting that Mitchell's size would help him stand up to Mr. G. He also suggested that in Bishkek we should talk to the Kyrgyz government about finding a replacement for Bozov. Mitchell was indeed much bigger than José, but his bulk was more middle-aged spread than muscle. He listened patiently as we told him stories about Bozov and Mr. G. His manner was fussy and mildly irritated. He took out a handkerchief and dabbed his perspiring forehead.

"Have you been to India?" he asked me. "I plan to spend a few days in Rajasthan on my way back to Washington." He perked up as we talked travel. But soon I steered the conversation back to the Aral Sea basin.

"Well, I don't know what can be done about these games," he said, sighing. "It seems that's the way everything's done here. You don't have much choice but to play along." He looked uninterestedly around the room and sipped the last of his coffee.

Wisely, José didn't bring up his idea to have Bozov replaced.

The next day Bozov arrived and we met with Valentina and her smiling Kyrgyz team. Bozov pretended that he and I were best buddies. We watched their new video, the story of a drop of water's journey from the Tian Shan Mountains all the way to the shrivelling Aral Sea. It was well produced, but it lacked any clear messages about saving water. Then Valentina presented her Kyrgyz public awareness strategy again, with her labyrinthine flow chart. José knitted his brow politely. Mitchell smiled aloofly and looked like he was thinking about Rajasthan.

"She is the best public awareness specialist in Central Asia!" announced Bozov when Valentina finished; it seemed she had overtaken Talbak again. Valentina frowned and shook her matronly head. Her team members were grinning proudly. So was Bozov, who added, "She's like a spring flower in our beautiful mountains!"

A few days later I was back in Tashkent sitting at the patio table at 45 Tazetdinova Street with four European men, all in jovial moods, happy to have escaped to this warm and exotic corner of the globe. Our consultants had arrived. Shakhlo was in her element, busy organizing things for them. She had hired the two Uzbek sisters, who were preparing meals, scrubbing floors and hanging laundry on the clothesline in the courtyard. Akhror, the owner, was filling the swimming pool and firing up the sauna. He was inviting us to try it out. Akhror's son, Alisher, had trimmed back the grapevine and was cutting the patch of lawn in the courtyard garden.

Frank Thevissen, dubbed Frank #1 as he was the first Frank to arrive, was our social research specialist. A bald chunky Flemish man of about forty, he was the director of a marketing firm and an associate professor at the Free University of Brussels. He was smoking a Gauloise as one of the sisters presented him with an espresso.

"Perrrfect," he said, smiling at her.

111

"Frank, we will go to Samarkand," said Frank Schwalba-Hoth – Frank #2 – to Frank #1.

Frank #2 was our public affairs specialist. He was a burly man with thick greying hair and beard and a mad grin. A founder of the German Green Party, he also lived in Brussels and headed his own consulting firm that specialized in lobbying the European Commission. He was wearing a beige safari suit, which I soon learned he always wore, its big zippered vest pockets stuffed with pens, books and notes and his ever-present camera. Frank #2 never stopped taking pictures. He excelled at social relations. He was a charming cajoler. I had experienced his talents in June of 1998 at the Environment for Europe Conference in Aarhus, Denmark, where he had astutely match-made guests as they arrived for the mayor's banquet. Surprisingly the Franks had never met before. But Frank #2's infectious grin and jolly disposition had already seduced Frank #1 and they'd immediately become friends. Frank #2 was reviewing Uzbek tourist sights, reading from his copy of *Lonely Planet Central Asia*.

"We will leave early Saturday for Samarkand. And on Sunday we will go on to Bukhara. It will be a weekend along the ancient Silk Road!"

"Perrrfect," said Frank #1. He rested his head back, closed his eyes and basked in the warm sunshine. "This is wonderful! In Brussels we haven't seen the sun in months."

Phil Malone, our TV production consultant, began giving the Franks travel suggestions. Having worked in Uzbekistan before, he'd already visited Samarkand and Bukhara. A robust and young-looking 40, he was a cheerful Liverpudlian with a store of corny jokes. When not in Uzbekistan, he was in Moscow, Sofia, Kiev or Tbilisi training teams of TV cameramen to produce videos. He had earned his stripes with the BBC.

Compared to the rest of us, Rick Flint looked positively youthful. Tall and lean, he frequently squinted and gaped as if what was going on around him was ironically puzzling. Originally from Leicester, England, he was a partner in a communications firm, also in Brussels, that worked on European Union projects.

"How is everything?" asked Shakhlo, coming out from the kitchen. She gave everyone a bold smile. "I try to make everything just like your home."

"Absolutely perrfect," said Frank #1. He sucked on his cigarette.

"Wonderful!" said Frank #2 with a grin. "Shakhlo, we need your help in arranging transport to Samarkand. What do you recommend? Plane? Train? Car?"

"My driver Faizillo take you," she said as if that were their only option. "I tell him. When you want to go? Saturday? It is possible."

When travel plans, sleeping arrangements and local transport were sorted out, I filled them in on how things were shaping up for the workshops.

"I think you should know that so far the level of cooperation has been a little less than ideal." I then related a few incidents to clarify what had been happening. "Hopefully the worst is over," I said with a smile. "And just to ensure everything runs smoothly, the World Bank has sent two monitors from Washington. They'll be sitting in on the workshops."

Frank #1 was unmoved; he had already worked in Kyrgyzstan. Frank #2 returned to browsing his *Lonely Planet*. But Phil was uneasy and Rick looked alarmed.

Mr. Rim Guiniyatullin rose slowly from his seat. His black suit jacket hung open and his dark tie rested on his massive belly. Directly over his head on the wall behind him was a framed photo of a bust of Uzbek President Islam Karimov; their haughty expressions were almost identical. Mr. G looked around the room, squinting at his audience. He fixed his stare on someone at the back of the long room. Then he put on his dark glasses and joined the tips of his fingers together, forming a dome. Nobody whispered a word.

It was the morning of the opening of the workshops. We were gathered in the Aral Sea project's large meeting room. The walls were lined with narrow pleated drapes the colour of boiled cabbage. The furniture – a long table, a podium off to one side that Mr. G had chosen not to use, several rows of lecture-theatre-type desks – was all matching light oak veneer. It was in a style reminiscent of the 1950s: spindly indented legs and clean lines. The padded chairs we were sitting on each had a white slipcover pulled snugly over the top, partly concealing the burgundy and salmon swirls of the upholstery fabric. The plastic casing over the sparse fluorescent lights in the ceiling had turned yellow with age, casting the room in an institutional murkiness.

Mr. G was standing at the end of a long table. Bozov sat to his right and Marina, a part-time interpreter who looked like a fashion model,

113

was to his left. Next to Marina were Paul Mitchell and José Bassat. I was next to Bozov and the rest of our consultants filled the remaining seats at the table. About forty participants, including the five national team leaders, sat at the desks in the back half of the room. Phil Malone had set up a tripod and was recording the proceedings with our new digital videocam.

Speaking in a low growl, Mr. G welcomed everyone. He acknowledged Paul Mitchell and José Bassat. He pronounced the names of our visiting foreign specialists. He talked about the national teams and their work. Marina, standing next to him, interpreted his words into English. They were Beauty and the Beast.

Then, just as I was settling into the stupor induced by his monotone, he snarled, "One of the foreign specialists and his assistant should be slapped across the face for the work they have done so far. They have destroyed the excellent work of the Uzbek national team!"

Paul Mitchell and José Bassat, across the table, locked eyes with me. José cocked his head as if to ask: *What is he saying*? *What did you do*? Bozov was staring at the table, half-hiding a happy frown. Phil poked his head around his camera and gawked at me.

Mr. G was referring to some misplaced newspaper articles and videos that belonged to the Uzbek national team. After reviewing them, Victor had returned them to Bozov. Then just a few days ago Bozov had sent one of his crazy memos, claiming that the materials were missing and Victor and I had destroyed them. Victor had reacted nonchalantly. "Look at the mess in his office. Probably Nadir lost them." We hadn't taken the accusation seriously.

"In these workshops we will evaluate their methodologies," continued Mr. G, referring to our consultants. "They will present them to us and we will choose the ones that we like, as if we are shopping at a supermarket. We will pick off the shelf those methodologies that we want to try and we'll see if they can work in Central Asia. If we don't like them, we will put them back on the shelf."

Then he stopped talking. He took off his dark glasses and glared at everyone in the room. There was an edgy silence for more than a minute.

"Now I am leaving," he said finally, "I know how I make you feel." A slight smirk emerged. "Now you can begin your seminars and get to work!"

He stepped away from the table. Then he lumbered past the table of consultants. When he reached Phil Malone's camera, he stared into the lens and growled, "Now we'll get them!" Then he continued down the length of the room past all the participants and out the door.

Four weeks later Jean-Charles Torrion would view Phil's videotape of the event and say, "Put it in the safe. It is *evidence*."

Two days after the workshop opened we were celebrating. Bakhtiyar Nazarov, the Uzbek team leader, had invited everyone to a Friday afternoon feast of *shashlyk, plov,* beer and vodka. We were in a private restaurant in the courtyard of an Uzbek home in a Tashkent *mahallah*. A crooner was wailing Uzbek songs to an electronic beat.

After the meal we started flapping around the courtyard. The Franks entertained a crowd of new fans despite language barriers. Phil floated around, videotaping the fun and cracking jokes. Even anxious Rick was smiling. Bayalimov shouted out anecdotes and toasts. I found myself dancing cheek to cheek with Valentina. She was giggling. It was a miracle; things had completely changed.

That morning Nazarov had announced that Frank Thevissen's workshop had been a success. This was despite the fact that most of the participants had never undertaken a poll before. And that Frank #1's marketing approaches were far more sophisticated than anything ever done in Central Asia. And that Nazarov hadn't even attended the seminar. This was all beside the point because Nazarov, the politician, was switching sides. It was a courageous move; he was inviting the wrath of Mr. G.

"Where's Bozov?" people began asking as the sun set. He was expected but had never shown up.

"Something must be up," said Victor, already suspicious of this sudden display of camaraderie. My mobile phone chirped.

"Rob, how are things going there?" It was Jean-Charles Torrion in Paris. He could hear the party going on and I explained that Frank Thevissen's workshop had gone well.

"Too well," I added. "It's a little fishy."

"Really?" I could tell by his voice that something was wrong. "Are you ready for this? Mr. G has just sent me a fax in terrible English – he doesn't even have a good translator! He says I must fire you and Victor Tsoy. Something about the Uzbek team's public awareness materials. That you both destroyed them?" He made a little disbelieving snort and let the news sink in.

My head was clearing. I moved outside the restaurant where it was quieter. Up the street I saw Faizillo sitting in Shakhlo's Lada. He stared

115

back at me. Victor had recently accused him of being a spy, along with Shakhlo and the woman who cleaned my flat, who turned out to be Shakhlo's mother. "It's ridiculous," I'd told him. "They're gathering evidence against you," he'd claimed. I walked farther down the road, away from Faizillo.

"Jean-Charles, this business with the Uzbek team's materials is complete bullshit! It's another one of Bozov's ruses. He's trying to frame us." I now felt completely sober. "But Jean-Charles, it's not Bozov. It's all coming from Mr. G. I'm sure he wants money. Shakhlo told me that's what he wants – and she should know. He's expecting a payoff."

JC was silent. Then: "Rob, he's trying to break the contract. Can you reach Anatoly Krutov right away?"

I walked farther down the road. The Uzbek music faded.

"Did you ever get a message that he wanted something?"

He hesitated, as if he was searching back through the six months of negotiations. *Why had they taken so long?* "No," he said finally.

"Well, I think that's the message he's giving us now. It's the only explanation that makes sense – the sabotage, the tricks, the attempts to undermine us at every turn. And now this trumped-up accusation that we destroyed these fucking materials."

After a short silent JC made a little laugh and sighed. "Rob, you have to talk to Anatoly. Let me know what he thinks we should do. But don't mention a payoff."

A few days later, late on a hot afternoon, the World Bank's chief of regional operations in Washington was sitting opposite me, dabbing his brow with his handkerchief. The Soviet air conditioner was chugging loudly but not generating much cool air.

"Do you know that he reads every night for four hours?" asked Paul Mitchell. He was talking about Mr. G. "He's read everything – all the classics, the philosophers, all the scientific research that's available in Russian." He seemed duly impressed.

The workshops were almost over. Anatoly Krutov, likely unable to deal with my situation, had sent his superior to advise me on what I should do. I had decided that I had three choices: cave in and fire Victor; quit and go home; or look into a payoff. I didn't like any of them.

"But he was extremely rude," continued Mitchell. "Marina was interpreting and she said he used the Russian word for 'fuck' several times right to my face." He turned back to me. "He's furious about your contract. He seems to feel that BDPA is not living up to his expectations." He was looking at me intensely. "But mostly he was complaining about Jean-Charles Torrion and Victor Tsoy. Especially Victor Tsoy."

"He wants something," I said. "I assume it's to break our contract. He never wanted it from the beginning."

Mitchell frowned sympathetically. "I think you have to do what he wants."

"If I fire Victor, then what's next?" Mitchell didn't answer. "It sends out a message that we'll stoop to his games. The games won't stop if I sack Victor. It's the wrong thing to do."

Mitchell mopped his brow again and sighed. "It's all very complicated. Central Asia is very complicated."

He looked out the window again for a minute, then turned back to me.

"You can rehabilitate yourself by removing Tsoy. Otherwise things will only get worse. You don't really have any choice."

You mean the World Bank isn't going to intervene, I wanted to say. But I didn't.

"It's their country," continued Mitchell. "It's their way of doing things. We're the outsiders here. We have to play by their rules." He eyed me grimly and then half-smiled. "I don't think Uzbekistan is any place to take a moral stand."

On May 30 Mr. G sent me a memo. He told me that he thought I should know that Canada was also suffering a water crisis. Canadians were dying from drinking their tap water. He'd learned about it on the internet.

I had no idea what he was referring to.

"Robert, he's only being sarcastic," said Ira.

Victor agreed. "He's only trying to irritate you. You should ignore it."

Later I learned what he was digging at. Seven people died as the result of an outbreak of E. coli in the drinking water in Walkerton, Ontario, in May 2000. The tragedy was blamed on provincial government cutbacks that had resulted in negligent water testing. It was the worst outbreak of E. coli in Canadian history.

12   Five Nations, Five Specialists

Soheil Ramanian ate his pizza meticulously with a knife and fork. In public school English he told me that his family had left Iran during the Islamic revolution in 1978. In London they were in the import-export business. Soon he was going to return to Tehran for the first time since he'd left. He smiled and said he was greatly looking forward to it.

We were sitting on the patio at the Bistro on Proletarskaya Street near the Fine Arts Museum. The Bistro was the current favourite lunch spot for Tashkent's tiny expat population. But on Wednesday, May 31, the place was mostly empty, just half a dozen expats and visiting consultants sitting under trellises crawling with grapevines. A paltry breeze barely fluttered the whisper-thin paper napkins in their plastic racks on the tables. It was 35 degrees Celsius.

JC Torrion had called from Paris and told me to meet with Ramanian. "He's done some work for us before, in the Caucasus – Georgia and Azerbaijan. He's also provided services for consultants in Tashkent. He knows the territory." JC gave a little laugh. "I guess he's your go-between."

As I ate my pizza with my fingers, I studied Ramanian's business card. He was director of a trade company with an address in Grosvenor Place, London SW1. I wondered how you got into the kickbacks-for-bosses business, what talents were required. Business savvy and discretion, I guessed. Excellent negotiation skills. (When I'd arrived for lunch, Ramanian had been closing a deal to sell *Cohibas* – Cuban cigars – to the Italian-Swiss owner of the restaurant. The owner had agreed to take several dozen.)

As I explained Bozov and Guiniyatullin's campaign against Victor and now me, I spotted David Pearce, the chief of the World Bank office in Uzbekistan, crossing the patio to join someone sitting in a corner. He discreetly ignored me. A few weeks before, in late April, JC and I had been sitting in Pearce's elegant office, admiring his collection of Central Asian *objets d'art* and carpets, relating much of the same story I was sharing now with Ramanian. Then Pearce had sounded sympathetic and reassuring:

"I think you should know that what's happening to you happens to all Western consultants here," he'd said in a mid-Atlantic accent. His tone was patronizing, but not unfriendly. "It's how outsiders have been treated in Central Asia for centuries." He'd smiled. "They say it takes six months before a consultant is allowed to start working here." Then more seriously, "But I'm becoming very concerned about the Aral Sea project. The region's water management is terrible and it's getting worse because of this continuing drought. Internationally this project's reputation is shit. And it's all because of Mr. Guiniyatullin. He refuses to cooperate. He makes everything very difficult." He'd stroked his white goatee and looked exasperated. "It's becoming impossible to work in Uzbekistan. There's a lack of cooperation at every level. Other donors are pulling out. The economy just keeps getting worse. It's going down the tubes!"

As we'd left, he'd promised to talk personally with Mr. Guiniyatullin. But Mr. G had played games with Pearce, avoiding him, then stubbornly refusing to discuss BDPA's contract. And Pearce had eventually told me that if things were still going badly by the end of June – six months into our contract – he would suspend Component B. But now as I watched him with his Colonel Sanders goatee, I realized that Pearce knew only too well how Mr. G operated and almost certainly that we were expected to pay up. Was he really going to do anything to help us? No, I thought, feeling snubbed. Payoffs were not something you discussed with the head of the World Bank in Uzbekistan.

I told Ramanian the latest twist in the game: Shakhlo had produced a letter allegedly from the American-funded NGO Counterpart Consortium, Victor's former employer, stating that they'd fired him the year before. It implied that he was a difficult personality and not honest. I was surprised and doubted the letter's authenticity. Victor could be obstinate at times, but he was scrupulously honest. I'd called his old boss, an American, and found out that no one at Counterpart Consortium had written the letter. It looked like Shakhlo had probably penned it herself and planted it in Victor's records in the Ministry of Foreign Affairs. She had been brazenly showing it around, infuriating Victor and stirring up gossip. I'd told her bluntly that Victor's past was none of her business and she was out of line. She'd said, "My friend at Foreign Affairs give me letter. I only give to Mr. Bozov because he is also my boss. It is my duty."

Ramanian listened attentively with his dark eyes fixed on me. We had finished our pizza and he was sipping tea. He was dressed in an expensive-looking suit and, despite the temperature, had not removed his jacket. The heat didn't seem to bother him.

119

"You know it's so much easier in Azerbaijan. There, they tell you right up front what they want. It's usually 15 percent of the overall project budget and you pay it off in neat monthly instalments. No hassles and everyone cooperates. But here, they don't tell you anything. You have to figure out what they want and if you don't, well you know the consequences." He smiled. "I can reach Mr. Guiniyatullin through the man in charge of the project's cars. The head driver. He's the contact. Seems a rather cagey fellow. I'll talk to him and have an answer for you by the end of the week."

Ramanian called me on my mobile a few days later.

"He seemed indifferent at first. But I knew he was interested. They're always interested. It went back and forth a few times. Then it was mentioned that Mr. Guiniyatullin had two sons at university in America and that $20,000 each might help sustain them."

"Forty thousand!" shouted JC through the phone from Paris. "This is too much! Mr. G wants to take away all of our profits!"

But $40,000 was just 6 percent of our contract budget. Of course there would be Ramanian's cut on top of that; still, according to the "15 percent standard" we were almost getting a deal! But JC refused to go the kickback route. "No negotiation," he said, and I was actually relieved. My guts, knotted for days, eased up. I realized that I couldn't bear to hand money over to the monster, one of the key instigators of the disaster we were supposed to be trying to alleviate.

That decision put me on the path to my first concession: to hire Bozov's "local specialists." According to the contract I was to hire five local experts: production assistants for TV, radio and print media types; an NGO liaison officer; and someone to set up and maintain a database. It all made sense in the contract, but Bozov had gone ahead and recruited three of these experts before I'd even arrived and he'd been demanding I hire them for months. Only one was qualified and I assumed they were all Bozie's Boys.

Downstairs Bozov took the news that I would hire "his" three specialists well. He lit a cigarette, eased back in his squeaky chair and leered at me.

"Robert! Finally you are cooperating. Finally you are listening to me. Now we can get to work."

The phone rang and he picked it up, tugging open his tie and babbling cheerfully. Nadir was gazing at me, aping Bozov's victorious leer. Then he rolled his eyes, grabbed one of Bozov's cigarettes and lit it.

"Tea?" he asked. "With *limon*?"

"*Kanyeshna*," I said. "Always with lemon."

But there was no lemon. There was never any lemon. It was our little joke.

The room was stifling. Bozov had told me air conditioners were "just for foreign specialists." A small fan hummed at the back of the room, barely pushing the stagnant air towards the open window. Their cigarette smoke wafted above our heads, bombarding a map of Kyrgyzstan on the wall over Bozov. Like Mr. G's office – in fact all the project offices – the walls were covered in maps. Maps of the five home states, maps of the Amu Darya and the Syr Darya watersheds. Maps of the Heavenly Mountains and the High Pamirs. Maps of Kyzyl-Orda province, the Autonomous Republic of Karakalpakstan and Tashauz province. Even Victor had decorated our office with maps of Uzbekistan and Tashkent. He too loved maps. But only Mr. G had a huge map of the Aral Sea. The smoke gradually drifted out the window, dissipating against the rusted metal blinds that partly blocked the scorching sun.

Nadir set down a bowl of tea in front of me. Bozov was still on the phone.

"Your only problem is Victor Tsoy," he confided in his loud whisper. He smirked. "It's not *you*! Mr. Bozov wants you to be his friend!" He laughed sarcastically. "It's Tsoy! Only Tsoy!"

"Nadir, what has Victor done?"

He shrugged. "You know that Mr. Guiniyatullin doesn't like him, and well ... ghhrrrrrrrrr." He made an ugly face and then revived his mocking grin.

I felt almost a kinship with Nadir, despite his silly carryings-on. I knew he had nothing personal against Victor or me; he was just a pawn in Mr. G's grand scheme. He understood well the role he had to play. But recently we'd learned that he wasn't as loyal as he appeared. In the process of recruiting our local specialists, Victor had suggested Nadir give us his CV. Of course Victor was only teasing him, but Nadir easily took the bait – the job would quadruple his salary. Then he'd pleaded with Victor not to tell Bozov. We were amused to discover that, according to his CV, Nadir was a *nuclear physicist*. Since then, whenever we discussed hiring our specialists, Victor would deadpan in his deep voice, "Robert, remember we need to have a nuclear physicist on our team."

"Robert ... Robert!" Bozov was off the phone. He regained his thoughts. "You will hire all five local specialists – it's already agreed!"

He claimed that each of the five countries had already approved its own specialist. This followed the project's base-five logic: five nations, five specialists. An Uzbek, a Kyrgyz, and a Kazakh that I had already agreed to, although they all lived in Tashkent, and now his last choices: a journalist named Boris Babaev, who must be the Turkmen specialist, and a Tajik that Bozov would soon decide on with Talbak Salimov in Dushanbe.

I swallowed the last of my tea and looked at them in mock bewilderment, as if I was drowning in the complexities of Central Asian politics and at a loss for words.

"Robert, you can't understand all this because it's our complicated way of doing things," said Bozov. He wagged his plump finger at me. "But you must do it! Don't forget that you nearly restarted the war in Tajikistan. Don't forget you destroyed the Uzbek team's materials. Don't forget that you stole all the excellent ideas of the national teams!"

"You have no choice," added Nadir, nodding accusingly. "You must hire all of them!"

They were both grinning. I grinned back.

"But Mr. Bozov, my understanding is that Boris Babaev is not a Turkmen," I said. "I would guess he's a Tashkent Russian." Bozov's base-five rationale had a hole in it.

"He's not!" Nadir rolled his eyes again. "He's Jewish!"

They both laughed. Then Bozov wagged his finger at me again. He leaned in close as if there were spies in the room. Nadir leaned in as well and translated.

"Robert! You *must* hire Boris Babaev. It will please Mr. Guiniyatullin. You must understand that Babaev is his friend. But Robert ... " His smoky breath was in my face. "If you also say goodbye to Victor Tsoy, *all* your problems will be solved!"

Then he sat back in his chair, eyed me slyly and picked his teeth with a toothpick.

"You are letting Bozov win," said Victor. "If you hire his specialists you will never have a chance to test your theories, try out your pilot projects and realize your dreams."

He was indignant and hurt and likely right. But I wasn't sure what he would have me do, and he didn't offer anything. Instead, he left abruptly for *Médecins Sans Frontières*. Which relieved me from having to tell him what was coming next.

A little later Sabit Madaliev was sitting opposite me at my desk. Sabit was a short, bearded Uzbek with a warm bucktoothed smile. He was a reputable Uzbek writer and had published several books of poetry. In the last years of the Soviet Union, he had lived in Moscow, editing a journal of Uzbek literature. But he'd returned to Tashkent in 1992, as Uzbek literature was no longer of much interest to Russians. In the mid-1990s he had tried out Uzbekistan's new private sector, working in the advertising departments at British American Tobacco and Daewoo Motors. These had not been good experiences.

"Terrible!" he said about the Koreans at Daewoo. He spoke excellent English and said everything with cheerful enthusiasm. "They were completely corrupt. Worse than Uzbeks." He smiled and said that he'd quit both jobs on points of principle.

Sabit had done some short contracts with the Aral Sea project during its first year, when there had been plans for the public awareness component to work with nongovernment organizations.

"But Mr. Guiniyatullin hates NGOs. So of course our attempts got nowhere."

Sabit was a breath of fresh air. He was completely different from Bozov's other specialists and I asked him how he'd made it onto Bozov's list.

"Your dear friend Bozie said that if Mr. Guiniyatullin interviewed me and approved me then I would get the job." Sabit laughed. "Mr. Guiniyatullin asked me if I believed in democracy and I told him, 'Yes!' Then he asked me if I thought Uzbekistan had a democratic government. I said, 'No! Of course not! And we have no freedom of the press or real human rights.' Mr. G had glared at me – you know that look of his." Sabit laughed again. "But he approved me anyway! Do you really want to know why? Do you want to know the truth? Of course it's not because I believe in democracy and human rights. It's because he knew that the other local specialists are terrible, really terrible. He needed one specialist who could really work." He thought this was hilarious.

Tulenbai Kurbanov, our new social research specialist, was in his late forties. He had droopy cheeks and the bleary eyes of an alcoholic. An ethnic Kyrgyz, he had worked at the Kyrgyz Embassy in Tashkent until "something happened" and he lost his job. Speaking some English, he told me that he'd suffered a heart attack a few months before and as a result he might miss a day of work now and then. He couldn't lift heavy things. Then turning sad, he said that his son had recently had an accident. Just 19, he'd hit a little girl while driving the family's sedan. He'd fled the scene. The little girl had died in hospital and the police had caught him and were now holding him. I told him I was very sorry. Tulenbai said he needed this job because he had to pay off the girl's family and the police so he could get his son out of jail.

"I will never drive that car again," he said in a maudlin voice. "The car is the killer, not my son!" He grasped my hand in his. "I'm going to sell that car and buy another one."

Our new public affairs specialist, Abbazbek Kasymbekov, was a nervous Kazakh in his mid-forties. He had a high-pitched voice that Sabit described as "pretty as a woman's." He spoke no English. He had once worked as a researcher at the Central Asian Scientific Research Institute for Irrigation, known by its Russian acronym, SANIIRI. In Soviet times its forerunner, where Mr. G had worked, had administered the regional water policies that had led to the Aral Sea disaster. The World Bank's Anatoly Krutov had also worked at SINIIRI, at the same time as Abbazbek.

"My God, Robert!" said Anatoly when I told him I'd hired Abbazbek. "I worked with him for three years. He was completely hopeless. He's an idiot!"

Boris Babaev, our new TV and video specialist, was a slim, fit-looking man in his mid-fifties. His wife and children had all moved to Israel but he told me that he belonged in Uzbekistan. He said he was too old to learn another language. He'd also never learned to use a computer. Yet he was much more astute than Tulenbai or Abbazbek. He was a seasoned reporter for Uzbek TV and knew how to put together a news segment. I'd watched some of his items on the evening news – it was fast-paced Soviet-style reporting with nothing controversial, but lots to please the bosses.

"I can help you," enthused Boris. "I will do a news story on your training. Everyone will see how good it is!"

He claimed Mr. G had promised him the job and a salary of US$800 a month. They all expected $800 a month, the amount on the budget line of our contract under "local specialists." I guessed that Bozov had been showing them this line, just as he had shown Shakhlo her budget line. The going rate for someone working on an international project in Tashkent with computer skills and fluent English was around $300 a month. Babaev's salary at Uzbek TV was about $50 a month. I started them off at $400.

I asked Boris if he would take leave from his duties at Uzbek TV for the duration of our project. He said that this wasn't necessary. He would keep this job "secret" from Uzbek TV.

"You can keep your job at Uzbek TV secret from me," I replied.

Jean-Charles Torrion flew into Tashkent from Kiev on June 17. He walked into the arrivals lounge in a cheerful mood, a boyish grin on his unshaved face.

"I am optimistic," he said as I took his bag. "In Kiev, we worked out all our differences with the Ukrainians. That project is now working very well. Now we'll fix this one!"

The road from Tashkent to Gazalkent took us along the Chirchik River past cotton fields, vineyards and apricot, cherry and apple orchards. Water gushed from irrigation canals into open conduits that crisscrossed the fields. Some spilled through cracks in the concrete, turning the parched ground a mottled brown. A heat haze blanketed the cotton fields, where the plants were just starting to reveal their fluff balls.

It was the last Saturday in June, and with the temperature over 40 degrees Celsius in Tashkent, our team was headed into the cooler Chatkal Mountains for a picnic and swim in the Charvak Reservoir.

Stands along the road sold *koumiss*, the fermented mares' milk that herders in the mountains produce each summer. Sabit and Abbazbek were in Tulenbai's new car, another Daewoo Nexia, the same model he'd disposed of because of his son's accident. They waved us over to enjoy a cup; JC and I were riding with Victor in his little Tico. JC sipped his gingerly. Victor abstained. In Mongolia I'd acquired the sour taste and got used to the gassy rumbling in my stomach. It now brought back memories of long treks in bouncing Russian jeeps across the Mongol steppe with frequent stops for *koumiss* at almost every yurt. Shakhlo and Ira were in Oksana's car; Oksana was one of our part-time interpreters. They made faces when I invited them to have a cup with us.

"Robert, women don't drink *koumiss*," said Ira bluntly.

From Gazalkent, the road climbed steeply into the mountains and the temperature dropped a few degrees. The scenery turned pastoral. A few herders on horseback led sheep and goats through grassy meadows.

The road twisted up through scrub and pockets of pine forest. We emerged onto a plateau that looked over a vast glacialgreen lake, the Charvak Reservoir. Yellow-hued stripes at the shore exposed the low waterline. It was a serious drought, the worst in forty years, said Victor, although the local media had hardly mentioned it. Water still filled the canals around Tashkent, the fountains and sprinklers still sprayed streams all over the city. Nobody was paying much attention to the drought. But farther upstream conditions were getting desperate. In the Autonomous Republic of Karakalpakstan on the remains of the Aral Sea, the cotton and rice crops were failing and the wells running dry.

We drove beside the reservoir, past President Islam Karimov's well-fortified *dacha*, his summer estate, to a suspension bridge that crossed a narrows at the far end of the reservoir. Just before the bridge we stopped abruptly at a roadblock. The Uzbek militia demanded written permission to pass, claiming that Wahhabis, members of a fundamentalist Islamic sect blamed for periodic anti-government violence, were hiding in the mountains. Wahhabis had been blamed for the terrorist attacks in Tashkent in February 1999.

Victor began to negotiate. After a few minutes he explained to JC and me that the town on the other side, called Nanoi, was Tajik even though we were still in Uzbekistan. Because Uzbeks distrusted the Tajiks, they set up roadblocks whenever there were any problems. Recently tensions had been escalating. After some animated talk Victor persuaded the militia to let him through on his own, and he went to find the mayor of the town – someone he knew – who would give us the written permission we needed.

An hour later JC, Sabit and I clambered down the steep bank, stripped off our clothes and plunged into the water.

"It's wonderful!" shouted Sabit. "Let's swim to the other side!"

Without waiting for our answer he swam off, easily beating us across the half-kilometre-wide reservoir.

We crawled up onto the packed-clay boulders and drip-dried in the sun. On the opposite shore smoke curled up out of the bushes where Tulenbai and Abbazbek were preparing *shashlyk*. The women were lounging and laughing, spots of bright colour blotched on a green pasture above the steep yellow bank. The sun, high overhead, blasted down on us. The rocky bank and the golden mountains rising above it reflected perfectly in the placid green water.

"It's a perfect day!" shouted Sabit. His happiness was like a boy's, exuberant and natural. It was rubbing off on us, easing the stress that had been building all week. Sabit dove in again and beckoned to us to join him. But we just laughed and watched him strike out across the reservoir again.

Despite JC's initial optimistic claim that we'd sort things out, the week had turned into a nightmare. My hiring of four of Bozov's local specialists hadn't stopped Bozov and Mr. G from refusing to approve our key specialist's mission. Patrick Worms was our star consultant. I'd known him since my Mongolia days and he'd got me the job as team leader and written BDPA's winning proposal. He was a busy executive with a PR firm in Brussels and was supposed to come and develop an overall public awareness strategy that was to guide all activities of the five national teams. Bozov and Mr. G's cagey refusal had raised the stakes considerably; our contract required we complete this strategy and win all five governments' approval for it. Of course, failing in our mission was exactly what Bozov and Mr. G wanted. But worse, in an attempt to win their consent for Patrick's mission we'd caved in to their demands and fired Victor, replacing him with Sabit. We'd suffered a huge moral defeat *plus* lost our bargaining position. And finally, when we'd tried to take our case to the World Bank, David Pearce was "too busy" to see us.

JC sighed and plopped a yellow stone into the reservoir.

"Victor seems to be taking the news okay," he said soberly. We watched the widening circle of ripples. After our week of horrors, the panorama around us seemed too spectacular and tranquil to be real.

"I think he's just relieved to be off the project."

We watched Sabit's spirited front crawl. He was approaching the far bank.

"With Patrick out, what can we do about this public awareness strategy?" asked JC as he tossed in another stone.

I chucked one at the same spot and we watched the ripples disappear. Then I looked at JC and snorted.

"Do you honestly believe they would ever approve any strategy we came up with now?" I laughed and threw in a series of stones, all at the same spot. "Well, with Sabit's help, we may have a tiny chance. But I can't imagine it ever getting approved by all five governments."

I grinned at JC. "I think they've really got us up a creek. And they know it."

127

We tossed in more stones in rapid succession. Then we stood and showered the reservoir with bigger and bigger rocks. It was as if we deliberately wanted to destroy the serenity of the place. After several minutes we ended our bombardment, looked at each other and laughed. Then we sat down and watched the scene turn idyllic again.

"I've worked on dozens of projects all over the world," said JC, "but never have I seen anything like this. *Never!* These guys are *really* awful! They do everything they can to stop us! But why?"

"I think you know why," I said.

"Okay," sighed JC. "So maybe I start looking for some money for Mr. G's sons."

He sounded angry. But when I looked at him to see if he was serious, he was grinning slyly.

"Rob, there will be a way out of this. There always is a way."

Sabit was on the opposite shore, signalling us to swim back. We waved back.

"Jean-Charles! Robert!" Shakhlo's shrill voice sang out over the reservoir, echoing off the mountains. "Please to come! Your dinner!"

13   The Walrus Club

On the table were two salads of chopped tomato, hot peppers and raw onion, a stack of warm round nan dotted with caraway seeds, two bowls of *plov* crowned with mutton chunks and a pot of black tea with lemon. Sabit closed his eyes, then brought his cupped hands together and passed them in front of his face; the *amin* is the Muslim grace, a gesture of thanks to Allah.

We were lunching at Sabit's favourite *chaikhana* on Abdulla Kodiri Street, not far from our office. It was crowded with Uzbek men in dark suits and women in print dresses and head scarves who worked for state agencies in the area: the customs police, the state telephone company, the state agency for registering artefacts, the state agency for nature protection. Vehicles roared by and the hot air was thick with fumes of gasoline, diesel and burning rubber. A stalled trolley was blocking traffic. The driver, a middle-aged Slavic woman, was yanking on ropes and trying to reattach the lines. She was unfazed; no doubt she'd done this a thousand times before. The trolley, an ancient model that looked like it should have been scrapped a decade ago, was jammed with people who stared blankly at us and watched us eat.

"Eat more bread," Sabit commanded, tearing off a chunk of nan and pushing it at me. He'd finished his meal quickly. Always restless, he operated at a high anxiety level. "You're not eating your salad. What's the matter? Aren't you hungry?" He filled up my cup with tea. Everything he said was a declaration, an order or an accusation. Yet the bucktoothed grin that followed these darts allayed much of his crustiness.

"Sabit, it's hard to eat the same thing every day. Tomorrow let's go back to Mafia Pizza." I smiled, saying this mostly to tease him. The restaurant I meant was a block away, on Navoi Street, and was actually called Fast Food. It was *my* favourite lunch spot. With its half-Westernized menu including *pitsa* and *gamburger* and its underworld ambience – shady business deals seemed to be going down – I'd renamed it Mafia Pizza.

"Robert, pizza is not good food." Sabit shook his head and clicked his tongue like a squirrel, a favourite Uzbek sign of disapproval. "And so expensive! Here you get an excellent meal for such a small price."

He was right; his *chaikhana* was great value. It was state-run and the prices were subsidized. It cost almost nothing to eat here. But I hankered after variety. Every day I suggested fried chicken, Greek, Indian or Turkish food, but Sabit kept demanding we come back here. I knew he was saving every dollar. He'd been unemployed for over a year before I hired him and he'd told me he must provide his two teenage sons with flats when they got married. His *dacha* in the suburbs also needed renovations. But when this job ended, he might be unemployed again for months, even years. "This economy is terrible! Prices always going up. Hardly any foreign investment. Hardly any foreign projects." Sabit was a great worrier.

"You must eat bread with your *plov*!" He ripped up the rest of the nan and plied me with another piece.

"You have very strict eating laws in Uzbekistan," I said, pushing the plate away.

Sabit shook his head, disappointed I hadn't finished my food. Waste! He jumped up and cleared away the dishes. Then we sipped our tea and contemplated our quandary.

"Terrible!" said Sabit, clicking his tongue again. "It's a scandal that they won't let Patrick Worms come to develop the public awareness strategy. Anatoly Krutov is not being honest with you. He should have told Guiniyatullin that Patrick Worms *had* to come. It's in your contract."

The day before, we'd visited Mr. G to ask his guidance on how we should proceed with the overall strategy. He'd seemed pleased that we'd come to talk to him. He bared his little yellow teeth. "This strategy will not be difficult," he'd said. "It should not be long. Only about six pages. Work with the Uzbek national team. Develop a strategy for them first. It will be the model. The other national teams will develop theirs from this one." It sounded like reasonable advice.

"What do you think Mr. G's planning this time?" I asked with a little smile.

"There's always a trick," said Sabit. He emptied the teapot into my cup and eyed me sharply. He was agitated. His eyes grew wider. "Robert! We can outsmart him. Yes! I know how to get the Uzbek national team on our side ..."

He began listing off every member of the Uzbek team, rating them on their understanding of public awareness, who would help us, who would be passive and indifferent, and who would only listen to Bozov and Mr. G. In five minutes he had it all figured out.

"Most of the team is completely useless!" He was nearly shouting. "They are only yes-men. Very weak! But they will do what everyone else does." He made this sound like wonderful news.

"What about Bakhtiyar Nazarov?" I asked. Despite the Uzbek team leader's questionable record with us, these days he was arguing with Bozov and showing us more goodwill. And he seemed to respect Sabit as a poet and a feisty Uzbek.

"Robert! We can do it!" His brain was churning so fast it was on the verge of exploding. "Nazarov will join us! I know how we can appeal to him. I have a plan!"

I looked at him skeptically.

"But Sabit, it's not just the Uzbek team we have to win over. It's Jalolov with his slicked-back hair and his ancient traditions. I told you he tried to sabotage our workshop with the Uzbek team back in March. He's on Mr. G's side."

"Maybe." Sabit was grinning. "But maybe not! You know what Bozie says – politics in Central Asia are too complicated for foreigners to understand!" He hooted. "Here is our plan: It's *their* strategy – the Uzbek team's – not BDPA's. Of course we'll write it, but *the Uzbek team* will ask Jalolov to approve it, not us. Once Jalolov approves it, Mr. G and Bozov will have to approve it because they must not upset the Uzbek government! You see?"

I saw, mostly, but it sounded risky.

"Remember what happened to me in Tajikistan."

"Robert! Everything in Central Asia is unpredictable. And dirty. It's all politics, tricks and lies." Sabit was grinning ridiculously. "Trust me, Robert! We will get this strategy approved!"

Over the next two weeks we drafted the Uzbek national public awareness strategy. We visited SINIIRI, the irrigation institute where Mr. G, Anatoly and Abbazbek had all worked, and collected documents on all regional water agreements signed since 1991. We pulled together any relevant quotes by President Karimov and Mr. G on regional water issues. Boris Babaev recorded a news segment for Uzbek TV on our work. He interviewed Mr. G, Bozov and me and the impression on the evening news one evening in early July was that there was Component B cooperation, even solidarity, on the developing Uzbek public awareness strategy. I liked this kind of propaganda. Boris was turning out to be useful.

These days Sabit was luring me into a new routine. Every day before lunch we left our office, crossed Navoi Street and walked past the soccer field, tennis courts and gym surrounding Pakhtakor ("Cotton-Picker") Stadium to the Ankhor Canal. This waterway provided the city with its only relief from the searing heat. With midday temperatures now reaching 45 degrees Celsius and never any rain, the canal was a major city attraction. Its glacial-green water rushed through the city, flowing under an arbour of willows, poplars and hemlock trees, past outdoor cafés, an outdoor concert bowl, a few neo-classical Stalinist buildings and some well-irrigated parks with cheerless monuments. Its shady banks were ideal for lounging and sleeping away the hot summer days. And its water was perfect for swimming.

Sabit was a long-time member of the *Murzh* or "Walrus" Club, a Soviet-style fitness club that operated all year on a bend in the canal. "Robert, for your health!" he'd announced as an inducement to join. But I was easy to persuade; I needed relief, not only from the heat but also from the stress generated by Bozov and Mr. G. Each day we greeted the old Uzbek men playing backgammon, pulled on swimming trunks in the dank change room and warmed up with a few rounds of badminton with the hardy, greying Slavs who strutted across the courts in skin-pinching swimsuits. Then we jogged a couple of kilometres along the canal to the Navoi Street bridge, beside the broccoli-green Ministry of Agriculture and Water, and dove in. The water was wonderfully clean and brisk. The current whirled us along and we crawled and paddled in it under the arbour, the leafy branches low enough that we could almost grab them. We whooshed past the high scaffolding surrounding the monument at Mustaqillik Maydoni, Independence Square. The Uzbek Independence Day celebrations were on August 31 and Karimov was already spending big money on them: a huge stage was going up for his invitation-only party. Around a bend in the canal we swirled into a pool of calm next to the Walrus Club dock. Here Sabit challenged me to an upstream competition. Catching our breath, we headed back into the current and swam against it, aiming to stay where we were or gain a few metres. Sabit always won, which delighted him.

One hot afternoon we spotted Uzbek team leader Nazarov on the dock at the club. He was standing proudly in his bathing trunks, looking robust and pleased with himself. We waved and he dove in and joined us.

"Does Mr. Bozov know where you are?" he asked me, laughing.

"Does the director of the Academy of Sciences know where *you* are?" I retorted.

On the dock we updated him on the progress of his strategy. Sabit lobbied him aggressively.

"It will be the model for the whole region!" he claimed with an infectious grin. "The other four countries will think of *you* when the use it to develop their *own* strategies. The Uzbek government will say that *they* developed their strategy first and *you* will get all the credit!"

Sabit the poet. Sabit the cajoler. Sabit the salesman. Nazarov couldn't help but smile. The plan was already starting to work.

On a hot morning in mid-July the Aral Sea project's cramped meeting room was jammed with 15 perspiring bodies. Under the photo of the five defiant presidents sat Bakhtiyar Nazarov and all seven members of the Uzbek team. Opposite them was our team, including Ira. A skeptical Bozov and Nadir sat at the head of the table. The Soviet air conditioner was humming loudly and Bozov ordered someone to pull the plug. In five minutes the room was stifling.

For the next hour Nazarov led a performance so bombastic that I was moved to silent gloating. Having suffered months of outrageous mudslinging and fabricated charges, I now sat back and let the members of the Uzbek team each stand and pontificate on the merits of the foreign specialist, his new Uzbek specialist and their outstanding contribution to their new strategy. We could do no wrong; we were above criticism. *Aha, Mr. Bozov! Look who now knows how to work in Central Asia!*

Bozov was outraged. He stared at the table, doodling on a scrap of paper during the overwrought speeches. Then he rose from his seat and glowered, nearly as effectively as Mr. G. Nadir, who'd been making quizzical faces at me, copied Bozov's expression.

"This is not the Uzbek national team's strategy!" roared Bozov. "This is Mr. Robert Ferguson and Mr. Sabit Madaliev's strategy! These ideas are not the ideas of the Uzbek national team. They are only the ideas of the foreign specialist. This is not a good strategy. It is not a good model for the whole region. It will never be accepted by the government of Uzbekistan!"

He sat down and started doodling again. Nadir looked at me and shrugged.

Then Nazarov was standing and railing. But not in Russian – in Uzbek. It was a tactic I'd seen him use once before, when he wanted to emphasize Uzbek solidarity. Bozov, who spoke Kyrgyz, could understand most of what he said. But Ira, sitting next to me, could translate nothing; like most Slavs born in the country, she spoke no Uzbek. In Soviet times, the Uzbek language had been denigrated to the status of a rural parlance of the uneducated. For Nazarov and Bozov's urban-educated generation, Russian was their first tongue and the language they were most comfortable with. But since 1991, the government had reinstated Uzbek as the official language, and the politician in Nazarov knew the effect he could make by speaking his "native" tongue.

It was pandemonium. For fifteen minutes everyone stood and shouted and waved and pointed. Except Ira and me. We watched submissively, squinting at the spectacle and feeling like aliens just realizing the huge mistake they'd made landing on this godforsaken planet.

"*Malchitye! Khvatit*! [Shut up! That's enough!]" yelled Nazarov finally, reverting to Russian. His voice boomed above the others and everyone quickly obeyed. He smiled at still-scowling red-faced Bozov.

"Many of us, including myself," said Nazarov calmly, "have been too busy to contribute what we could to this strategy. It is very good, but now we will make it excellent! We will rewrite it right now! We will include *all* of our ideas!"

The nodding spread rapidly around the room. Then poor old Bozov had no choice. He stood up and gave his angry wave of assent. Then he mumbled something profane and stormed out of the room. Nadir scampered after him.

Three weeks later I was walking from my flat to the office under the trees that shaded the wide park in the centre of Abdulla Kodiry Street I had just returned from my mid-contract holiday in Europe. Water rushed through the open gutters nourishing the giant poplars that swayed and creaked in the hot breeze. One tree had recently crashed to the ground; over-irrigation had saturated the soil, allowing the wind to blow it over. Several people were hacking off branches for their *shashlyk* barbecues.

"Robert!" Sabit was suddenly in front of me. He threw up his arms. "I was coming to see you." His grin was as wide as his face. "How could you have left me! You have no idea how terrible these people are. I have spent the last three weeks killing myself. Look at me! I'm mostly dead."

"Sabit! You don't look very dead to me. But you do look even crazier than when I left."

We hugged each other and sat down on a bench. Dog walkers, young women with babies and out-of-work men strolled past, indifferent to two men having an animated conversation in English.

"We have rewritten the strategy a thousand times!" said Sabit, growing even more excited. "But don't worry it is still not really changed. And you won't believe it! Nazarov and all the Uzbek team members have signed it. And Robert! Just a couple of days ago Nazarov showed it to Bozie. And Bozie went crazy! They yelled at each other. Nazarov called Bozie an idiot to his face! Robert, can you imagine?" Sabit hooted.

"Poor old Bozie!" I laughed. "I would love to have seen that!"

"And that useless Nadir! He was gawking at Nazarov like he was shocked, but then *he* burst out laughing. Then Nazarov and I started laughing. Poor old Bozie! He looked so scared and miserable, like when Mr. G yells at him. I almost felt sorry for him. Robert, he lost his few little brain cells and couldn't say a word!"

"But what about Jalolov? Has Nazarov talked to him yet? Is he ever going to sign it and make it official?"

"Robert, Nazarov says he is too busy." Sabit shook his head and laughed. "Too busy swimming in the canal! But I have seen Jalolov three times. Each time he says the strategy is good. But he won't sign it. I keep asking him what is wrong, what can I do to fix it? But he won't tell me. Then at our last meeting – just yesterday – I showed him the signatures of all the members of the Uzbek team. Then I gave him a book of my poems! I could see that he was impressed. Then I said, 'Now, *tell* me what you want!' I was firm with him. Finally he said that the strategy should say more about our traditional approach to water use. He wants even more of that. Robert! Even more! So I am writing a *preface* to the strategy. I mention our long history of efficient and prudent irrigation. I mention Uzbekistan's greatest poets and writers. I mention how the Soviets destroyed our traditions. Then I quote President Islam Karimov. Robert, it is pure genius! He will love it. He will have no choice!"

"I hope your holiday was very nice, Robert." Shakhlo was cool. She updated me on routine office matters. Then she said, "Something very serious happen. While you have your very nice holiday someone steal

$400 from safe." She fixed her bold gaze on me. I didn't say anything. "It was Victor Tsoy," she said resolutely. "Only he has key to safe. You give him key. I see you."

"Shakhlo, Victor returned the key to me. And anyway I never told him the combination. It can't be him. You and I are the only people with the key and the combination."

"Victor is calling me every day," she continued, ignoring me. "He say he want money and I must pay. I say only Robert can pay. He must wait for you. But he was very rude. And I know he is coming here, looking at my things." She opened the drawer of her desk and looked inside to demonstrate.

Under Victor's new agreement with BDPA – JC was keeping him on as BDPA's regional rep – I did owe Victor his July salary, which JC had transferred to my account. But $400 also happened to be the exact amount of money Shakhlo needed to top up her salary to the contract budget line. She was not only probably upping her wages once more, but also trying to frame Victor again – killing two birds with one stone. I almost laughed out loud at her audacity.

"Robert, Shakhlo is Uzbek," said Ira when I asked her privately what she knew about the missing cash. She looked hard at me and stiffened. Then she shrugged and tried to look indifferent. "I would never trust an Uzbek."

"Did Victor come here while I was away?"

"Of course Victor didn't take the money!" She was annoyed. "Who do you think is telling Shakhlo to do this? Robert, is it ever going to stop?" She was almost in tears. "Don't you see? There's only one way to stop it!"

She quickly recovered and turned away from me.

One evening a few days later I was sitting with Shakhlo on the couch in my living room. There was a problem with the electrical system and my air conditioner wasn't working. Faizillo was tinkering with the circuit box.

"It is problem with all flats now," said Shakhlo unsympathetically. "Electricity is weak. You must not use air conditioner."

"But Shakhlo," I laughed. "It's 45 degrees! Besides, I have no electricity right now."

She shrugged. She and Faizillo had driven me over in her new purple Lada. I knew that a new Lada cost US$3500 and I was wondering how she could afford it, whether she was dipping into project funds again. Her belligerence was getting worse.

"Can you please ask your mother to keep the door locked when she's cleaning the flat?" I asked her. "The other day I came home and found the door wide open. Several neighbourhood kids were sitting here watching TV. My camera and other things were lying around. They could have been stolen."

"Robert, I tell you! Please. She is *Luba*." She gave me a meaningful smile. She was still insisting that my Uzbek cleaner was a woman named Luba and that "her friend in Moscow" owned the flat. When I'd told Sabit this, he'd hooted. "But Luba is a *Russian* name! And of course her mother is Uzbek! No Uzbek has a Russian name! It's a stupid coverup. She doesn't want you to tell anyone that her mother is your cleaner because everyone will know she's getting even more money from you. And the flat of course belongs to her mother!"

"Luba is not happy with you," chastised Shakhlo. She pointed at the balcony and the blood-red Turkmen carpet I'd bought from Aziza in Samarkand. "She say you are walking on carpet on balcony with shoes. This is very bad. No shoes in this place."

"I'm very sorry," I said with a little sarcasm. "But after all it is my carpet." She gave me a chastising look. "Of course it won't happen again."

The TV and air conditioner suddenly started working. Shakhlo smiled.

"There! You see, now everything is fine."

Bozov was holding his stomach and scowling. He looked miserable He ordered Nadir to make tea. Then he lit a cigarette and blew the smoke into my face.

"Robert, the Uzbek national strategy is no good. You cannot use it as a model for strategies with the four other national teams. *Again* you are not listening to me!"

"Again you are not listening to Mr. Bozov!" echoed Nadir.

Abdurakhim Jalolov, the pomaded Uzbek national coordinator, had finally signed the document. Sabit's preface had done the trick. It was hard now not to glory in our triumph, but I was trying hard to show some humility. After all, I reasoned, I hadn't come to Central Asia to gloat.

Nadir set a cup of tea in front of me and shook his head disparagingly.

"Mr. Guiniyatullin is very angry at you," continued Bozov. "You did not consult him before Mr. Jalolov signed this document. Again you are making big problems in Central Asia."

"*Very* big problems in Central Asia!" Nadir shook his head at me and scowled. "*Very* big problems for Mr. Guiniyatullin!"

I sipped my tea and pretended to be repentant.

"I'm *very* sorry Mr. Guiniyatullin is unhappy. Of course I did go to him about the strategy. But he advised me to show it only to you. He said that you are the boss of the public awareness component, not him. I only did what I was told."

"Robert, I want to help you to understand us," said Tulenbai. He was very earnest. There was a side to our social research specialist that wanted very much to please me. "I will tell you a story that will help you to understand."

It was a hot Saturday night in mid-August and we were celebrating the approval of the Uzbek strategy with a party on the patio at our house on Tazetdinova Street. All the members of the Uzbek team had come as well as our team, including Ira and Shakhlo. But not Bozov and Nadir. There had been many speeches and toasts. Now everyone was very drunk.

"In the early 1980s I won a scholarship to study in America." The drunker Tulenbai got, the more his English improved. "When I arrived, I was amazed. It was very different from Kyrgyzstan. Everywhere I saw things that I never saw at home – big cars and shops with wonderful things. And the people were not cold and miserable and unhappy. They were friendly. I did not expect that."

He poured more vodka into our tall glasses. We were sitting at one end of the table, which was littered with beer, vodka and wine bottles, plates with leftover *shashlyk* and cake and ashtrays full of butts. Uzbek pop songs were blasting out the small stereo that Shakhlo had rigged up on a window ledge. A few people were fluttering, spinning and staggering around the courtyard. Others were in small groups, talking and laughing.

"But then a strange thing happened. I got sick. For many weeks I couldn't do anything. I couldn't go to classes or study. I just lay in my bed. Then one day I realized that everything they'd told me about America was a lie. It was not an evil place. The people were kind and good. They didn't hate me. Then I knew that I didn't believe what they'd told me. And after that I started to get better. I went back to my classes. Soon I was fine again. But I realized that something had changed. Do you know what had happened?"

I thought for a few seconds. I hadn't a clue. "You got used to hamburgers?"

"I lost my communism."

I laughed. "Tulenbai, that's a terrific story!" I held up my glass. "Let's drink to lost communism!"

Tulenbai smiled weakly and we emptied our glasses. Despite his confession, his conversion wasn't very convincing. Not much about Tulenbai was very convincing.

"Robert," he pleaded. "I only want you to be my friend."

"To friendship!" I shouted as Tulenbai refilled our glasses. "To lost communism and new friendships!" I stood up and the table and the patio swayed beneath me. I held my vodka high and announced loudly to everyone: "To capitalism! To the new Uzbek national public awareness strategy! To Aburakhim Jalolov! To Kadirbek Bozov! And of course, to our great leader, Mr. Rim Guiniyatullin!"

Then I looked down at unsmiling Tulenbai.

"Tell me, my friend, does 'Rim' really mean 'Rome' in Russian? Is Mr. G really named after that decadent empire?"

"Please, Robert," said Tulenbai, eyeing me uneasily. He was still holding up his full glass. But he had nothing to worry about; no one else was listening to me. "You must understand that you must be friends with Mr. Bozov. It must be your success. This is most important. More than national strategies."

"More important than rescuing the Aral Sea?"

"Robert!" shouted Sabit, rushing up beside me and pulling me back into my chair. "They are saying $50 is not enough. They are all so greedy!"

His eyes were bulging. He hadn't touched a drop of alcohol but he still looked crazier than the rest of us. He was talking about the crisp US$50 bill I'd given each member of the Uzbek team. Sabit had disapproved. "You're bribing them!" he'd scolded. "No, Sabit," I'd rationalized. "Bribes are paid before. I never promised them anything. This is a bonus for signing." But Sabit had shaken his head and clicked his tongue. "Robert, sometimes you are a terrible liar! If you give them money there will be trouble."

"They are demanding $500 each for the strategy. I told you! I told you it was a big mistake to pay them!"

"Sabit, you're exaggerating. They're too drunk to care."

"Robert! It's terrible! Don't you see how bad they've become? All of them. Even your good friend Nazarov. He's the one who told me. He's behind it!"

I scanned the courtyard and found the Uzbek team leader entertaining a group with a swaggering speech.

"No, Sabit. It's not Nazarov. It's poor old Bozie! He told them I must pay them. I can hear him – 'Fifty dollars is not enough! It's an insult! Ferguson must pay each of you $500! After all, he's stealing all your excellent ideas!'"

I laughed and turned to Tulenbai. But he looked very disappointed with me.

14   The Return of the Franks

We found the driver standing behind his Ford Transit van smoking a cigarette. A middle-aged Slav, he had clear blue eyes and was smartly dressed. As Shakhlo had said, he looked reliable: "Robert, it is very good van. Red! And driver is Ukrainian. Ukrainian is excellent driver!"

The Franks were returning for their second round of training and I'd asked Shakhlo to find a van to take them to Kyzyl-Orda to work with the Kazakh team. But I thought the price, $150 a day, was high. And when, as with the house rent, she'd shown little interest in negotiating it down, I'd said to Sabit with a little smile, "Let's undertake a little detective work." Sabit had been delighted to comply.

Thanks to our driver, Sabirjan, we'd tracked the van to this garage. The driver seemed more anxious about the logistics and hazards of the proposed trip, particularly the notorious Kazakh *militsiya*, than the price. Almost indifferently, he told us the rate for the van was $120 a day.

"She was going to take $30 a day for herself!" shouted Sabit on the way back to the office. "She's a liar and a thief." He was outraged and thrilled. "Robert! She is terrible!"

But I was more surprised at her sloppiness than her dishonesty.

"That was pretty easy. Now we'll talk to Akhror and see how much she's skimming off the house." Akhror was the owner of our house on Tazetdinova Street. I smiled at Sabit. "What I need is some hard evidence."

But I'd already made my decision: Shakhlo had to go.

Sufism is a path, a journey. A destination. Along the path, the Sufi acquires knowledge of "reality." God is the ultimate reality. As one book explains it, "The aim of Sufism is the elimination of all veils between the individual and God."

Sufism is the mystical side of Islam. It involves meditation, trance and revelations in a personal search for God. According to Abu Nasr, a classical Sufi teacher who died in 990 CE, "The Sufis are people who prefer God to everything and God prefers them to everything else." Like most religious beliefs, Sufism isn't lacking presumption.

Sufism has a long and successful history in Central Asia. Its focus on the individual and tolerance for other beliefs made it particularly amenable to the free-spirited Kazakhs and Kyrgyz, and it's credited with seeding Islam in the region around the year 700. It functioned through *tariqas*, sects or brotherhoods, led by *ishans*, teachers or sheiks who provided spiritual guidance. Holy places, often tombs of the *tariqas'* founders, served as focal points. "Sufi" means both "pure" and "wool," after the wool coats worn by early followers. Sufis are also called dervishes, from the Persian word *darvesh* meaning "poor." Dervishes used to go from door to door begging for food and lodging. Their twirling, trance-like states gave them the name "whirling dervishes."

In early May, Gabi and I had visited a Sufi holy place outside Bukhara. Near the village of Kasri Orifon is the tomb of Bakhautdin Naqshband. In the 14th century he founded one of the most important orders of the Sufis. He had been a sort of Islamic Gandhi, teaching the importance of a full spiritual life while maintaining a serene harmony with the world. The Naqshbandis are non-fanatical and secretive. They are known for their practice of *zihr*, chanting the names of God and sacred verses while holding a strict posture and controlling their breathing.

A huge dome covered the *khanaka*, the Sufi contemplation hall. Beside it was a slightly tilted minaret and a courtyard with two restored mosques. Naqshband's tomb was a two-metre-high marble slab in the courtyard. Nearby was another slab, this one draped in sheets of colourfully striped silk – purple, vermilion, lime green, Egyptian blue – like the bolt Gabi had bought. It was shaded by a mulberry tree with a trunk half a metre thick.

A young mullah, his eyes closed, was sitting on his feet in the midst of this sea of colour, dressed in a blue-grey robe and a *dopy*, a four-sided black skullcap. He was young, about twenty, with fine, clear features. His expression was composed and faraway. He was practising his *zihr*, but he wasn't audibly chanting any words. A teacup, a bowl and a red plastic basin lay in front of him. The basin was for donations and its bottom was covered in *sum* notes and coins. A crowd had gathered, gazing at him in admiration. The men wore robes and *dopys* and the women were draped in long dappled dresses and shawls. Gabi and I were drawn to him as well. He had an ethereal tranquility that made it seem as if he were floating. We were at the mausoleum for more than an hour and he never moved once.

Muslims endured great persecution during the Soviet era, but Sufism survived because it did not rely on mosques for public prayer. Despite the KGB's efforts to infiltrate it with spies and crush the anti-Communist *tariqas*, Sufism not only secretly endured, but provided a network for an anti-Soviet underground movement. It has never shied away from politics. Since the end of the Soviet era, Central Asia's reviving Islamic culture has had a strong Sufi flavour, and has won many more adherents. It's also providing a peaceful alternative to the Islamic fundamentalists, who despise it.

Sabit was a Sufi. Every day after lunch he returned to his flat and meditated for an hour. When he returned to the office he was calmer, refreshed. The panic in his eyes had subsided. He was more philosophical. For Sabit, Sufism was a sort of salvation, a release. After decades of confronting Soviet autocrats – and these days Mr. G – he'd acquired a conspiratorial view of the world that was now firmly entrenched. He struggled daily to come to terms with the world's good and bad sides. Too often the bad side seemed to be winning.

"Robert, the world is a terrible place," he often said, shaking his head and clicking his tongue.

Sufism also supplies the believer with a moral code.

"When the day of resurrection comes, God will examine everyone's life," he might say humbly. "God will examine the balance of good and evil that every person has collected during their life."

These days I knew that he was thinking mostly about Shakhlo.

On a hot Sunday afternoon in late August, Shakhlo parked her purple Lada inside the gate at 45 Tazetdinova Street. She pulled the garden hose across the garden and through the basil hedge and started washing down her new car. Frank #1 was smoking a Gauloise and watching her. She giggled and sprayed him. He tossed away his damp cigarette and chased her. She screamed. She laughed. Frank sprayed her, and her soaked shirt outlined her nipples and her belly dancer's tummy.

Sabit and I went to find Akhror, who lived a block away. He was out but his wife said he would come and see us when he returned. He showed up around seven, just as we were about to eat Shakhlo's *plov*. All of our team was there, sitting around the table on the patio. We were celebrating the return of the Franks.

"You don't like my *plov*?" asked Shakhlo suspiciously as Sabit and I excused ourselves for a few minutes.

Robert Ferguson

On Tazetdinova Street I told Akhror a little lie. I said that BDPA in Paris required a receipt from him for the amount he had received for each month's rent. We had receipts already, but Shakhlo had written them all. I said a note in Russian was fine. Sabit had brought along some paper and Akhror wrote that he'd received $900 a month, not the $1,200 that Shakhlo had claimed, for each of four months. That added up to $1,200 for Shakhlo.

"Robert!" Sabit was shocked and delighted. He could hardly contain himself.

Akhror realized something was up and became alarmed. I explained to him that he had just confirmed that Shakhlo was not being honest with us. With the news out, Akhror pounced on her: he said that Shakhlo often called him, complaining that things were not right at the house.

"She was scaring my wife with terrible accusations! I tried to do what she said, but she was never happy. She kept saying that Robert is angry with me because I don't make the house right. Maybe Robert will change to another house!"

I tucked the evidence into my pocket and we returned to finish our dinner. Shakhlo's eyes pierced me as I sat down. She and Frank #1 were sharing a large bowl of *plov,* eating in traditional Uzbek style, stuffing rice into each other's mouths with their fingers. They were enjoying themselves. Everyone else was eating theirs with forks. Sabit and I sat down and Sabit performed his *amin.* Shakhlo started laughing, a snicker that grew into a shrill cackle. She held up her glass half full of red Uzbek wine.

"We should all be very thankful to our team leader for arranging such a nice party." Her sarcasm was biting. Everyone hesitated before drinking their wine. There was an awkward silence. Then Frank #2 stood up and grinned.

"If I may, please?" He deferred to me and I nodded. The moment called for diplomacy, Frank #2's forte. "Here's to our wonderful team leader, Robert, without whom none of us would be here." He bowed his head to me. "And here's also to Shakhlo, our most extraordinary office manager, who is a charming cooker of *plov*, this remarkable dish, which is the national food of wonderful Uzbekistan." He bowed his head in her direction. She giggled and looked at Frank #1. "And I must also make some most kind words for our most distinguished Frank Thevissen, or Frank #1 as he is now known famously all over Central Asia. He is also very famous in Brussels, where he is now advising the

144

prime minister of Belgium on polls that show what the Belgian people are together thinking." There was a chorus of "ooohs." The two Franks bowed their heads slightly to each other and grinned.

"And also we have here two very clever men, Tulenbai and Abbazbek. Very lucky are Frank #1 and me, Frank #2, to be able to work with both of them, who will train us to be experts in everything in Central Asia." More grins, bows and giggles. Tulenbai and Abbazbek looked very pleased. "And finally we have two of the most charming of all the Russian women in all the former Soviet Union. We have beautiful Ira, our most talented interpreter, who makes every day much more brighter because she is always in our office." Ira blushed. "And also the most lovely Oksana, our second but not second-choice interpreter, who makes me now wish only to live forever in wonderful Tashkent." Ira and Oksana gawked at each other and laughed. "To all of them and all of us!"

We all stood and toasted.

The next morning I fired Shakhlo.

"Robert, something has happened," she announced calmly as she walked into my office. Her hair was pulled back and knotted in a scarf and she was wearing a flowered summer dress. She looked very self-assured. "Someone has come again to the office. You know I tell you before that we have thief. Please come look safe."

I followed her into her office. Ira was sitting at her desk working on a translation. Behind Shakhlo's desk, the safe door was wide open. Ira stood up and excused herself.

"Please," said Shakhlo, pointing. I looked in the safe and found a pile of receipts, a balance statement and the original version of the Uzbek national strategy. But no Uzbek *sum* or U.S. dollars. I looked at Shakhlo. "You see. Somebody take everything. I tell you. We have thief."

I was amazed by what she'd done. Again I couldn't believe her audacity. Did she really think that she could get away with it?

"Shakhlo, I've said this before – only you and I have the combination and key. Maybe you gave the combination to someone else and loaned them your key." My tone was slightly ironic.

"I tell you. Sometimes Victor come here. Other people tell me they see him. He is thief. Everyone know." Her voice was shrill but her demeanour unruffled.

"How much money is missing?"

She sat at her desk, silent for a few minutes as she went through her records. She appeared to be adding up the missing cash. I assumed she must know the amount she had taken already, but maybe not. Anyway she had to carry on with her charade. I watched her, expecting her to fall out of character and break into her bewitching smile. Then she would hand over the cash and we would have a good laugh. *You see, Robert. Everything is fine.* Conceding defeat, she might even admit that Bozov had put her up to it. Or possibly Mr. G. And if that really happened I would forgive her. Or would I?

Then a second wave of thoughts hit me: How could I get the money back? Bozov would say that he didn't believe me. Mr. G would accuse me of being responsible for the theft, not Shakhlo. Maybe her move was not as foolish as it seemed. Maybe I was being set up again.

She looked up from her desk.

"In *sum,* about 18,000. In dollars, I think 1,200."

With her take on the rent, that meant she'd grabbed a total of $2,400. Likely she'd skimmed a bit more here and there – how else could she have afforded the new Lada? Still, in project terms it wasn't big money. It wasn't going to upset JC. I recalled his words back in April: "It's usual to fire the office manager after six months. It's the only way to stop the pilfering." Then he'd laughed. My concern now was to avoid a scene, to prevent Bozov from turning this into a scandal.

"Shakhlo, I'll make a deal with you. If you return the money that's missing from the safe, you can leave freely and I won't mention any of this – including the money you have been taking on the house, which I know all about – to Mr. Bozov or Mr. Guiniyatullin. And I won't call the police."

She stared at me for a few long seconds, seemingly dumbfounded, but probably considering my offer.

"You think *I* take money?" She blinked slowly and fixed her gaze on me. "You think *I* am thief?" Her voice was shrill. "I cannot return money because *I* not steal money." Then she sneered. "You ask Victor for money!" she shouted. She threw back her head. "Police! You tell police about money *Victor Tsoy* take from you!"

"Shakhlo, I have lots of evidence on you. I have a receipt from Akhror that proves you were taking $300 each month on the rent. I know that you were planning to take $30 a day for six days – another $180 – on the red van. If you don't return the money I will take you off the project."

I repeated my offer again, sweetening it a little: I said that the Franks would never know anything about it. Or Jean-Charles Torrion, although I doubted I could really keep it from him – he enjoyed any juicy gossip I produced about Shakhlo. But she was obstinate and unrepentant. She wasn't going to back down.

"*I* not take money! *I* am not thief!"

"How much money is in your bag?" I asked.

"You want me to pay you *my* money? I have almost nothing." Her eyes flaming, she emptied her bag out on the desk. There was a mobile phone that I hadn't seen before. From a wallet she pulled out a $50 bill and some *sum*. "I have daughter, I must pay for her school." She waved the $50 bill. "I must pay this to headmaster! His extra fee or my daughter cannot go to good school! I have many expenses! You don't understand! You don't care about me!"

"Shakhlo, I only want you to return project money that you should never have taken. Stealing money is not going to solve your problems."

She stared coldly at me. She counted out her cash in front of me. But the stolen cash wasn't there and she wasn't going to tell me where she'd put it. She'd probably passed it to Faizillo earlier, who'd driven off and locked it in a safe somewhere.

I asked her to leave the building. Slowly she collected her things. Her mobile phone rang. In a subdued voice in English she said she'd call back. Probably Frank #1, I thought.

I walked her to the elevator. She got in and stared hard back at me.

"Now you have *much* bigger trouble," she declared defiantly as the doors closed.

"The aim is to discredit BDPA and me," I said to Sabit. "Theory 1: Bozov ordered her to do it, but Mr. G sanctioned it. Everything they do always comes down to him. Or rather goes up to him. It's all part of his campaign to discredit our work and show the World Bank that he doesn't need foreign specialists. He hates public awareness, he hates Western public awareness specialists and he hates our contract because the World Bank forced it on him. Mostly he hates the World Bank, he wants to punish *them*, to show them that we are not only unnecessary, but incompetent. His specialists are better than our specialists. Foreigners get lost."

Sabit shook his head and clicked his tongue. He was lapping it up.

147

"Or, Theory 2: Good old greed. Mr. G just wants money." (I hadn't told Sabit about the $40,000 fee for Mr. G's sons as I knew he'd be disgusted with me for having taken it as far as I did: "You were negotiating a bribe?" I could hear him rebuke.) "Pay him off and he's our friend. Welcome to Central Asia."

"Robert! He is so terrible, he is the devil!" Sabit's eyes were crazy. But then he grinned. He knew that his mad denunciations delighted me.

"Or: Bozov is behind everything. Theory 3: Bozov alone is setting me up. He sabotaged our workshops. He's the one who told Shakhlo to write the letter framing Victor. He's the one who destroyed the Uzbek team's materials and then blamed Victor and me. He ordered Shakhlo to clean out the safe. Bozov reports what's happening to Mr. G and Mr. G blames me because Bozov is convincing him that it is me."

"Impossible!" shouted Sabit. "Old Bozie can't write a memo! Old Bozie can't understand the Uzbek national strategy! Old Bozie can't think!" He grinned again after that outburst. Calling Bozov an idiot gave him great satisfaction.

"Or," I said with a grin, "Theory 4: Shakhlo acted on her own greedy impulse. She knew the gig was up. She knew that I knew that she was taking money. So she grabbed all she could. Then she tried the rather lame excuse that Victor Tsoy did it. It doesn't seem very imaginative. But it's consistent. And now she will rally her forces against me. And they may be considerable."

"She won't get away with it!"

"She just might," I said. "I have a feeling she just might."

"Never!" shouted Sabit. "She will pay for this!"

"You have no real evidence that she stole this money. You are blaming her, but it probably was someone else." So rationalized Frank #2.

It was early afternoon now and the Franks were sitting opposite me in my office. Sabit was meditating at home. Their gloomy, accusatory faces told me that Shakhlo had given them her side of the story. (I could hear Shakhlo, that unrelenting look on her face, her big eyes enraged, telling the Franks, "Robert think I steal money! He think I am thief! But it is Victor, Sabit and Robert. Together they steal money. It is them!")

"Who?" I asked.

148

"Victor Tsoy," said Frank #2. But he was grasping. I knew he didn't really believe it was Victor. He liked Victor. "It could be anyone in your office."

"Victor is gone and he never had access to the safe. And Ira? Do you really think so?" He didn't answer. "Frank, only Shakhlo and I have the key and the combination to that safe."

"Then it's Sabit," said Frank #2. "He's a nasty little man. I don't trust him."

Frank's jolly charm was gone. *He* sounded nasty.

"Sabit is possibly the most honest Uzbek in the country. Besides, he had no access either."

Frank #2 got up. He'd had enough. He revived his jovial old smile as he walked out the door. Frank #1 was eerily quiet. He sucked on a Gauloise and looked at me like I was not only wrong, but also heartless.

"She's a single mother," he said, finally breaking his silence. "She's just got divorced." This was news to me. "She has two children. Her older boy is disabled. Do you know how difficult it is for a woman like Shakhlo in a country like Uzbekistan? They don't respect women here, especially if they are ambitious and strong like Shakhlo."

"Frank," I sighed, "I have other evidence against her." I told him about the house rent and the red van. "I can't work with an office manager who pilfers money. I know it's not huge amounts, but I have to have someone trustworthy looking after our accounts."

"But you're completely wrong," he scoffed. "Shakhlo did not steal the money from the safe. You need someone to blame and it's easiest to blame her. If she took a little money from the project here and there, that's not so serious. Everyone in this country does this. It's a poor country and people have to survive."

We stared at each other for a few long seconds. Frank, slumped in a chair, exhaled his cigarette smoke and looked magnanimous, morally superior, a forgiver of small transgressions. On the side of the underdog. I was the mean and petty one, picking on society's victims, failing to see the big picture and missing the point.

She's bewitched him, I thought. He's completely and stupidly in love.

"It's not such a serious matter," said Frank #1. "Why are you making it into such a big deal?"

"I agree, it shouldn't be a big deal. That's why I made her an offer." I explained it to him. "I don't want the police involved, or Bozov and Mr. G. But if she doesn't accept this, things will only get more complicated, and much worse for all of us. Believe me."

149

Frank #1 studied me as he thought it over. Then he got up to leave.

"I'll talk to Shakhlo," he said despondently. "But you are completely wrong about her. She didn't steal your money from the safe."

Bozov was in a gloomy mood. So was Nadir.

"Robert, you must write me a letter with your evidence. If you have strong evidence, then I must accept it."

The news didn't seem to surprise him; his reaction was subdued. He knew all about it, I thought. Theory 3: He put her up to it. And now that she's done her job and been caught, she's dispensable. But what was in it for him? A little extra cash? The delight in humiliating the foreign specialist? But now he's lost his number one spy.

I told him that Mila Islamova, one of our part-time translators whom he knew, would start the next day as our new office manager. Anticipating Shakhlo's dismissal, I'd approached Mila about the job the week before. Bozov eyed me suspiciously, as if this was too quick work on my part. But he nodded and raised no objections. He looked defeated.

"You and the Franks will go to Ashkhabad on Thursday," he announced. "It's all arranged. Take their passports to the Turkmen Embassy tomorrow."

As I left his office, I thought: No, it's not poor old Bozie. Sabit is right. Not enough going on upstairs. Back to Theory 1: It's Mr. G. The fat bald man has been baring his tiny yellow teeth and pulling everybody's strings. He's playing with all of us, and thoroughly enjoying himself.

An old Uzbek woman was trudging up a lane between some apartment blocks. She was carrying a big striped canvas bag, obviously empty. She looked angry.

"Look!" I said to Sabit. "There goes 'Luba.' She must have been at my flat, but couldn't get in. But what's she doing? It's not her cleaning day."

It was lunchtime the following day and Sabit and I were sitting at his favourite *chaikhana* on Abdulla Kodiry Street, which was not far from my flat. We had finished our *plov* and were sipping lemon tea.

That morning I'd had the locks changed at my flat. Shakhlo's mother, "Luba," had access to it, which meant that Shakhlo could also get in, and I was worried that she might try to get revenge. I didn't want to return home to find the flat ransacked and my things missing. I also realized, unenthusiastically, that I would have to move. I couldn't continue to live in Shakhlo's "friend's flat."

"Let's ask her what she's doing!" Sabit's eyes were twinkling.

Before I could stop him he had jumped up and rushed over to her. I joined them in the midst of a heated exchange in Uzbek. After a few minutes Sabit summarized it for me:

"She's very angry with you for changing the lock. She says she must get in and look at her things to see that they are okay. You must change the lock back again. And it was very bad of you to fire her daughter. Before, Shakhlo worked with Americans and they were very good men. But you are a terrible Canadian. You are not like them. You are the worst man she has ever met!"

He grinned at me, then turned back to "Luba" with a look that was both respectful and skeptical.

I agreed to let the furious old woman into the flat and we walked together to it. Once inside, she went directly to the storage space under the balcony. When she saw me looking at her, she shooed me away.

In the living room a few minutes later she chastised me again in Uzbek. To appease her, I told her through Sabit that I was moving out right away and that I wanted to pay her for her cleaning for the month of August. I asked her how much she wanted, in dollars. She dismissed the question but I insisted. Finally she said that I must ask Shakhlo.

"I'll bet Shakhlo's ripping off her own mother," I said when she'd left. "You can bet she's not paying 'Luba' the $80 that I give her for cleaning. And I think I know where the stolen money is." I showed him the safe under the balcony. "Why else was she so upset about me changing the lock? Probably she just took out the money right now and it's in that striped bag of hers. We should run after her and get it back."

"Robert!" shouted Sabit. He made a move to rush after her out the door.

But I pulled him back and laughed. I wasn't desperate enough to chase after old Uzbek women and grab their bags. Not quite yet.

151

15   The Battle of Bishkek

Far away in the house the doorbell cooed, its sound the mournful call of a wild pigeon. It was almost 3:00 a.m. and I was lying in a narrow, sagging bed in the guesthouse next to the sauna. The servants' quarters. I had moved into 45 Tazetdinova Street and the Franks and I were now roommates. After a minute of woeful cooing, I heard the kitchen door creak open and a pair of feet in slippers shuffle across the patio stones.

"Frank! Is that you?" It was Frank #1 calling to Frank #2 in a loud whisper.

"Of course it's me." Frank #1's weary voice. "I can't unlock this fucking gate."

"But already I showed you!"

The deadbolt squeaked and the gate creaked open.

"Frank, you must put the key in just so. Not too far. Like this." The key squeaked back and forth in the lock. "You see? You must push it in like this. Now you try!"

Silence. Then the jiggling of the key in the lock. Then a sigh. Then another squeak and the gate clicked closed. Then a tired laugh from Frank #1.

"I think you see it now," said Frank #2 brightly.

"I can't get it," said Frank #1, exasperated. "I'm going to bed."

The next night the same thing happened again.

In the days following Shakhlo's firing, Frank #1 was rarely at 45 Tazetdinova Street, or the project office. He was spending all his time with Shakhlo. That left Frank #2 and me most evenings sitting at the patio table looking at each other. We didn't have a lot to talk about. He read and reread his *Lonely Planet Central Asia* and made cheery observations on anything that popped into his head. His philosophy was a potpourri of positive thinking, statistics he'd picked up – mostly the huge volumes of water required to grow cotton and rice – and his trademark buffoonery. He was congenial and amusing. He was glib and trying. Like a tireless salesman, he never stopped with his spiels.

"The world is full of great coincidences," he said one night, showing me a picture he'd taken at a conference several years before in Almaty, Kazakhstan. "Look, it's Mr. Bozov!" Bozov, his arm slung over the shoulders of the well-known leader of a local NGO, looked haughtily exasperated. Both men appeared uncomfortable. Bozov had no love for NGOs. "It means we met each other before. Isn't it wonderful!"

"How do you always maintain your upbeat mood?" I asked, careful not to sound flippant. I had only witnessed his mask fall once – when he'd defended Shakhlo – and I suspected that like most hyper-happy people, his suppressed alter ego could be vicious.

He dropped his manic grin and for a few minutes spoke honestly. He told me that when he was a young man in his twenties, he reached a point where life was losing meaning for him. He couldn't find satisfaction or happiness in anything he did. He'd even considered suicide. Then he'd had a revelation: if he was going to survive, he had to find a way to exist and be happy; he had to find a new role. So he stopped being an earnest environmentalist (I imagined the early version of Frank as a '60s radical, a righteous and committed shit-disturber) and reinvented himself as a merry public affairs consultant. He joked, he cajoled, he flattered. He took pictures, he remembered names, he looked for coincidences. He gave everyone his mad-happy grin. And it worked.

The night before our departure for Ashkhabad, the other Frank and I were sitting alone at the patio table avoiding each other's eyes. We didn't have a lot to talk about. Frank #1 was smoking a Gauloise and looking happily exhausted. The lack of sleep was catching up to him, but he was glowing. I had no doubt that he was in love with Shakhlo.

"It's been very difficult," he said. "She's still so upset, so shattered. She can't believe what's happened to her. It's as if she's living in a dream, a dream after a terrible nightmare."

I knew he was trying to win my sympathy, to get me to forgive and forgo my wrong-minded accusation. Back down and let bygones be bygones. It was an international project. Things like this often happened. I was only trying to use Shakhlo as an example. She was the victim. I was the one looking bad here.

I pictured them sipping red wine and chewing *shashlyk* in outdoor cafés, gazing at the Ankhor Canal and then at each other, sighing, smiling and blinking away sudden wells of emotion. Then back to her flat in the purple Lada for some lovemaking. The kids were in bed asleep, they wouldn't hear. (According to project gossip, she'd just got a divorce from her husband, who Frank #2 said was a layabout. He also claimed it had been a loveless marriage.)

"It was *so* very difficult, but in the end she wrote it." Frank #1 handed me her letter of resignation. He said that he would pay off the missing money through a deduction in his pay from BDPA. "Rob, you are completely wrong about her. I know that she is innocent. She took nothing from the safe."

After he'd gone to bed, I relayed his offer to Jean-Charles Torrion in Paris.

"She cleaned out the safe? She blamed Victor? She was stealing money on the house? She is *unbelievable!*" JC laughed. "I never met anyone like Shakhlo before."

"There's more. She and Frank #1 are having an affair. He believes she's innocent. He admits she might have skimmed the house money but he says that everyone does that here so it's all right. He's completely bewitched by her. She's been filling his head with stories – you can just imagine."

"Incredible!" JC laughed again. "Never a dull day in Tashkent. How are you surviving all this?"

"I pinch myself each morning to see if I'm still really here. Unfortunately it seems that I am."

Ashkhabad was an oven. The surreal city that had glittered in dancing snow last March was now a surreal city basking under the cruel Karakum sun. Turkmenbashi's palaces, mosques and statues shimmered, rippled and wilted. His fountains sprayed recklessly in the heat haze. His pines and palms, swollen with Karakum Canal water, swayed and flapped in the blasts of heat exhaled by the Kopet Dag Mountains. His people trudged the streets in languid dazes.

After the drama of the past few days, Ashkhabad was a great relief. A respite. It turned the Franks silly. Frank #2 set the pace with relentless comedy routines, and Frank #1, with a Gauloise always going, couldn't stop tittering at everything Frank #2 said.

The Franks and I checked into President Niyazov's son's luxurious Nissa Hotel, managed by Italians and full of Italian contractors keen on some fun but unable to find any. We spent the afternoon with Usman Saparov and his new Turkmen team, an informal gathering of polite old men in a hot room. The team was passively attentive, but there were few questions. They had yet to start their work. A huge poster of Niyazov was propped against one wall: Turkmenbashi, always watching.

The next day Usman took us on a "cultural outing." We drove east from Ashkhabad down the newly twinned and blacktopped M-37 highway to Geok-Tepe, site of the Russian massacre of 15,000 Turkmen in 1881. President Saparmurat Niyazov had erected a huge mosque with several azure domes and minarets on the site. Usman claimed that it was the largest mosque in the world. We gazed at it for an hour from the highway as one of our jeeps had broken down. In Bakharden we descended 60 metres below the rocky ground to an underground lake, Kov-Ata. We skinny-dipped in its 36-degree-Celsius mineral waters that smelled of rotten eggs. Usman guaranteed us its healing powers would extend our lives. The Franks were delighted with this news.

Later we had a picnic in a shaded spot beside a mountain stream in the Kopet Dag Mountains, only a few kilometres from Iran. Lunch was half a roasted sheep, Turkmen nan and Turkmen vodka. After the meal we dunked ourselves in the gushing water to drown the heat and alcohol. Then, dozing under a tree, I got the feeling I was being watched. I opened my eyes and squinted into the sun. On top of a rocky peak about 200 metres away was a familiar squirrelly face. It was another poster of President Niyazov, out here in the middle of nowhere: Turkmenbashi, always watching.

At Ashkhabad airport Frank #2 wore his new Turkmen outfit: an ankle-length quilted coat in radiant blue that looked like a dressing gown and a black sheepskin *telpek* hat that was a good match for a '60s Afro. An Air India flight from London to Delhi was on a stopover and the waiting area was full of bored Indians. Frank #2 strolled between the rows of seats, hand over his heart, bowing and saying: "*Asalam aleykum! Asalam aleykum! Asalam aleykum!*" The passengers just stared at him. A child burst into tears. Frank #1 and I watched from the safety of the bar. Frank #1 couldn't stop laughing.

"Robert! You missed my birthday party. But we didn't forget you! We drank to your health!" Dauletyar Bayalimov laughed. Then his glazed eyes searched for the two Franks who were sitting across the table from him, grinning inanely. They looked as hungover as Bayalimov.

"Ohhh, Frank," moaned Bayalimov; this seemed to be directed at Frank #1. Then he dropped his head onto his folded arms and in a minute was snoring.

155

It was Saturday morning, a week after the Franks and I had returned from Ashkhabad. We were in an office at the university in Chimkent, Kazakhstan, with several water management officials from southwest Kazakhstan – Chimkent's province, which with Kyzyl-Orda province made up the Kazakh share of the Aral Sea basin. Sabirjan had driven Sabit, Ira and me here from Tashkent that morning to rendezvous with the Franks and Bayalimov. The Franks and Nadir had just spent four days in Kyzyl-Orda with the Kazakh team.

The representatives from southwest Kazakhstan were polite and inquisitive. They didn't know anything about the Aral Sea project. They had not been included in any of Bayalimov's activities. Bayalimov snored throughout the meeting.

"I've been throwing up for three days," said Frank #1 during tea break. "Bayalimov kept coming into my hotel room and forcing me to drink vodka." He laughed but he looked awful.

"It's Bayalimov's vodka terrorism," I explained. "Very effective. You don't stand a chance. All you can do is stay as far away from him as possible."

"Frank, I told you already my system," said Frank #2 with jolly authority. "You pour it in your water glass. You dump it on the floor. You put it into flowerpots. You toss it into the air after his toasts. Do anything but drink it!" Then he took me aside. "Team leader, we must get back to Tashkent now! We're flying to Bukhara this afternoon."

Frank #1 looked so miserable I doubted he'd make it, but I sent them back with Sabirjan. The rest of us would return later in the red Ford Transit van.

Late that afternoon, despite my advice to the Franks, I found myself face to face with Bayalimov in a bar with rough timber walls, swinging half doors and Kazakh waitresses dressed as cowgirls. Kenny Rogers and Dolly Parton were singing a duet.

"Robert," slurred Bayalimov after chugging down his mug of beer.

"Do you know that Frank Thevissen is the very best public awareness specialist in the world?"

"Yes, I know that now," I answered. "But did you know that Valentina Kasymova is the very best public awareness specialist in Central Asia?" He studied me for a few seconds. "And that she is as beautiful as a flower in the Kyrgyz mountains?"

I poured half my beer into his mug, which he then held up.

"Robert!" he shouted. He didn't allow even a hint of a smile "To our excellent public awareness specialists! And to those two bastards Bozov and Guiniyatullin!"

Issyk-Kul, Kyrgyz for "warm lake," sits 1,600 metres above sea level in a huge dent between ranges of the 4,000-metre-high Alatau Mountains. Its lightly salted, clear water is thermally heated and never freezes. Issyk-Kul is the second-largest alpine lake in the world. In Soviet times it was designated Central Asia's alpine Riviera: spas once lined its sandy shores and the Soviet elite dropped in for mudpacks, to "take the waters" and to acquire a Kyrgyzian tan. Like the Aral Sea in the 1960s, vacationers came from all over the USSR to bask on its beaches. It was a sort of Soviet Lake Geneva, a Kyrgyz Baden-Baden.

And like the Aral Sea, its waters were also under threat. Kumtor, a Kyrgyz/Canadian-run gold mine operating in the area, had recently admitted responsibility for dumping more than two tonnes of sodium cyanide into the Barskoon River, one of the lake's feeders, in 1998. The environmental and political fallout is still being felt. As well, Valentina claimed that the lake level was starting to drop, the result of drought and overuse of the water destined for the lake. She suggested that our next project should be a new public awareness project to save Issyk-Kul. She was serious.

A week after our rendezvous with Bayalimov in Chimkent, Frank #2, Sabit, Ira, and I were on the beach at Lake Issyk-Kul (Frank #1 was back in Europe for a month). With us were Valentina and her four smiling Kyrgyz team members. We had just completed a two-day workshop in Bishkek. Things had gone rather well. We hadn't achieved anything, but we weren't lambasted in the local press and nobody had stood up and condemned us. To celebrate I had invited everyone for a weekend at the beach.

But Issyk-Kul's beaches were abandoned and its hotels and spas empty. It was September 9, and at this elevation summer was rapidly succumbing to a brief, sharp autumn – it would be snowing here in just a few weeks. But few vacationers were coming any more even in midsummer as Kyrgyzstan was now demanding they pay in U.S. dollars.

We huddled on the deserted beach. Heavy clouds hung low over the lake and the wind carried a chill off the hovering mountains. Vodka bottles appeared, followed by Valentina's matronly smile of consent, which impelled Sabit to lead me promptly into the tepid water. His

Sufism had led to a strong disapproval of drinking, any drinking. He was in an anxious state, tired of workshops that accomplished nothing, bored with Frank #2's jokes and missing our swims in the Ankhor Canal. He was a bundle of repressed energy.

"Let's swim to the other shore," I proposed.

"Robert! It's 70 kilometres!"

"Is that a problem?"

Thrilled, he charged ahead to a buoy about half a kilometre out, easily beating me and reviving his boyish enthusiasm for life. That was as far as we got.

Back on the beach, the afternoon sun came out, turning the lake turquoise, almost like the ocean in the tropics. The clouds lifted and the mountains, freshly showered, glistened like wet slate. Water droplets hung in the air and a rainbow appeared, much to Valentina's delight. The sun gave the place a little cachet, some of the glamour of a real resort, and for a minute I squinted and saw gleaming skin and bikinis sprawled among dozens of striped parasols *à la St-Tropez.* Radios blasted Russian bubble-gum pop. Out on the water were splashes, gleeful shouts and laughter, and a traffic jam of paddleboats. Across the sea of roasting bodies was a row of towering Lombardy poplars that shivered in the breeze. And beyond them nothing but the steep treeless mountains. The vast desolation all around reminded me that we were in the middle of nowhere. And when my eyes returned to the beach it was just as empty, except for the lonely parasols pitched in the yellow sand.

Sabit made us a huge and satisfying *plov* for dinner. Uzbek men traditionally cook this dish, much as the barbecue is a male domain in Western suburbs. After dinner, the food, alcohol and holiday spirit closed the day on a note of comrade-like affection. And triggered an inspiration: "Let's hold our upcoming regional workshop right here, at Issyk-Kul! It will be perfect, even if it's snowing."

Everyone was excited by the idea. Except Valentina.

"Robert!" she warned, destroying the agreeable ambience. "You can't decide anything without Mr. Bozov's approval."

"Not Issyk-Kul!" said Bozov. "You will hold it in Osh!" Back in Tashkent, Bozov and Nadir were eyeing me slyly. The location of the regional workshop was our new game.

But Osh was an odd choice. In Central Asia, Kyrgyzstan's second-largest city was synonymous with ethnic violence. In the densely populated Fergana

Valley, Osh was dominated by Uzbeks; it was only a few kilometres on the wrong side of the Uzbek border thanks to one of Stalin's notorious divide-and-rule scrawls. During three infamous nights in 1990, hundreds of Uzbeks and Kyrgyz were massacred near the city while the Soviet militia stood by, idly watching. Incidents and rumours of clashes still abounded. It was a dangerous place, a Central Asian flashpoint. And it would be nearly impossible to get everyone visas.

"It's all arranged!" Bozov explained that the year 2000 was Osh's 3,000th birthday – city elders claimed the city was older than Rome – and a huge patriotic celebration was planned for October. Hundreds of officials from the Fergana Valley were going to descend on the city to attend meetings and celebrate the anniversary. "You will hold your workshops at the same time!" he said, smirking.

A few days later I found out why: his draft budget for our workshops included all the officials' parties; he expected BDPA to pick up the tab for the whole event. I had to get our workshop out of Osh.

Sabit and I spent most of September 18 searching Tashkent for Canadian whisky. It seemed impossible to find. Finally the bartender at the Intercontinental Hotel agreed to sell his backup bottle of Canadian Club. His price was ridiculous, but it was for a very worthy cause – it was Mr. Guiniyatullin's birthday.

At the end of the day Mr. G studied his bottle and bared his small yellow teeth. He thanked me and said that we should drink it together sometime.

"How about right now?" I suggested. Getting pissed with Mr. G, I thought. What a wonderfully dangerous idea. It might finally break down our barriers.

But he just glared at me and then I was dismissed.

"Where are those two bastards?" shouted Dauletyar Bayalimov. "Where are Bozov and Guiniyatullin?"

Bayalimov and his Kazakh team had missed the opening of our regional workshop, choosing instead to lock themselves in their hotel room and drink. When they'd finally shown up, Bayalimov had started interrupting our sessions with frequent catcalls: "Why aren't they here? They *deserve* to be here! If we have to be here, they *must* be here too!" His impish grin soon had everyone laughing, even Valentina.

It was October 2 and we were in Bishkek. The Osh conspiracy had collapsed. The celebrations had been postponed, then the regional conference cancelled. Politics had got in the way. But Bozov, exasperated, had refused to concede defeat. "Robert! Valentina has booked the beautiful Hotel Issyk-Kul. Next door is President Askar Akaev's palace. You'll love it!" But the Hotel Issyk-Kul turned out to be in Bishkek. It was a Brezhnevian concrete hulk, still state-run, built originally for visiting high-level Communist apparatchiks. Its faded Soviet grandeur complemented its bad Soviet-style service.

Bayalimov's taunts were making it clear why we were all in Bishkek: this workshop was to be a battle. A few days before I'd left Tashkent, Sabit and I had run into Nazarov, the Uzbek team leader, swimming in the Ankhor Canal. He'd apologized for not being able to come to Bishkek. "Robert, I must warn you," he'd said. "Mr. Bozov and Mr. Guiniyatullin are very angry with you and me for our strategy. I think that you will have big trouble in Bishkek. Please be careful." Then Bozov had also chosen to stay away, leaving Lieutenant Nadir to rally his troops.

We grouped into our factions and prepared for battle. Bayalimov, Valentina, Talbak, Tulenbai Abbazbek and Nadir used favourite Bozov techniques: scowls, denouncements and complaints. Talbak was the chief whiner. The food was awful, he griped. Why was I holding the workshop in this terrible hotel? He demanded I give him money so that he could go to a restaurant for a decent meal. I told him that Valentina had made the arrangements for us all. But Valentina followed up with grumbles of her own. Ira filtered most of these for me without even translating them.

Our side – Sabit, the members of the Uzbek team, and Phil Malone and his video crew – pushed on with the agenda despite the resistance. We smiled, to keep up morale and to rile the other side. Many of the participants were caught in the middle, trying to figure out what the hell was going on. A few, like Frank #1, were trying hard to play on both teams.

Everyone was wondering how things would end. Usman Saparov, leader of the Turkmen team, pulled off a feat that surprised us all. He stood up and announced:

"As leader of the Turkmen team and technical director of the Aral Sea project, I propose that the local specialists of the BDPA team – Sabit, Tulenbai, Abbazbek and Boris – prepare a report summarizing all the activities, ideas and recommendations that came from the participants of this workshop. This report will be sent to everyone as the official assessment of its success."

It was brilliant. No one dared to reject his suggestion. It seemed to make sense. Minutes later, Sabit pointed out how this was going to play out.

"Robert! The others can't write the report. They don't know how. Only I will write it!" His toothy grin was from ear to ear. "Just wait till poor old Bozie reads it!"

The snow on the mountain trail was deep, tiring our horses. Gabi Buettner, my Bukhara travel companion, was ahead of me, her more experienced horsemanship giving her an edge on the slippery terrain. Behind me was Bruno de Cordier of Ghent, struggling with his stubborn horse. He yelled joyful curses and now and then pulled off his hat and whooped like a cowboy. Anything American amused Bruno to no end.

We came up onto the flat top of a small mountain and stopped for a rest. We dismounted and let the horses paw the snow, searching for a weed to chew on. The wide Chui Valley spread out below. Above us were the snow-covered Alatau Mountains. A white yurt with smoke rising from its peaked felt roof stood 500 metres off, almost invisible in the snowy meadow. The sky was starting to clear, but it was cold. Winter up here already. It was October 7.

The hike in the mountains was Gabi's idea. She loved horses and it was a way to celebrate her thirtieth birthday. Bruno also worked for UNDP Kyrgyzstan, coordinating their information and providing support to the resident representative. He was built like a rugby player and his eyes were wild with a playful sort of lunacy. As we looked at the view, he told us a joke in his throaty Belgian-accented English:

"A Kyrgyz shepherd is watching his flock when he suddenly sees a bright red, lavishly chromed Jeep Cherokee racing up the hills in a large cloud of dust. With shrieking brakes, the vehicle comes to a halt right next to him. A slick young foreigner with Ray-Ban sunglasses jumps out, puts his Armani shirt and Brioni jacket right and asks the shepherd, 'Tell me, dear man! If I guess the exact number of sheep you have here, will you give me one?'

"The shepherd looks at him for a few seconds and then says calmly, 'Fine.'

"Then the young foreigner pulls out his laptop computer, connects it to his cellular phone, gets into the internet and downloads the NASA site, scans the region with his GPS satellite navigation system, activates a database with 60 Excel tables, prints a 150-page report with his portable high-tech printer, turns to the Kyrgyz and says, 'You have exactly 1,586 sheep over here.'

"'Indeed,' says the shepherd. 'So now take a sheep.'

"The young man takes one and puts it in the trunk of his Cherokee. Then the shepherd asks, with an amused look in his eyes, 'If I guess your profession, will you give me the sheep back?'

"'Oh well, why not?' says the man. "'You are an international business and management consultant.'

"'Right,' says the surprised foreigner. 'How do you know that?'

"'Very simple,' the shepherd replies. 'First, you come all the way up here while nobody ever asked you. Second, you ask for payment to tell me things that I already know. And third, you don't know anything about who we are and how we live. And now give me my dog back!'"

16   Water Is for Fighting!

The ancient Uzbek in the foyer was gone. Every morning for the past eight months he'd been there, steadying himself with his security podium and acknowledging everyone with an indiscriminate nod and a mumbled *"Asalam aleykum."* He'd been a fixture. But the day after my return from Bishkek, I found he'd been replaced by five Uzbek *militsiya* officers in olive-green uniforms and frying-pan hats. Their weapons were slung over their shoulders and their grim expressions declared that the building was now under military control. One of them rifled through my passport searching for my visa. From now on, he said, I must have a special pass to enter the building. The Aral Sea project must issue me one right away.

"The KGB are ordering us out – IFAS, the project, all the private companies – everyone!" said Mila upstairs. She was upset. Our new office manager was sitting on the windowsill blowing her cigarette smoke out into the chilly mid-October morning and chewing nervously on her thumb. "Robert, what we will do? Why I didn't stay as interpreter? It is much better job." The daily responsibilities of bookkeeping and dealing with cagey project staff were already wearing her down.

The Aral Sea project occupied the seventh and eighth floors of a building belonging to the Uzbek KGB, and they were reclaiming their space. Since the terrorist bombings in Tashkent in February 1999, which the Uzbek president had pinned on an opposition group of fundamentalist extremists linked to the Taliban in Afghanistan, the military had been rounding up suspects tied to radical groups. Apparently President Karimov was stepping up these operations and expanding the KGB. Already about 15,000 men, many just innocent religious Muslims, were locked up in the new gulag called the Jaslyk Penal Colony near the Aral Sea. "The prison of no return," Sabit called it with paranoia in his eyes. His Sufi beliefs and beard made him a potential inmate.

Late that morning two young conscripts burst into my office. Without saying a word they ordered Sabit and I out of our chairs. They pulled the desks from the walls and, using blowtorches, began cutting out the old radiators. The noise and stench were terrible.

"They never worked anyway," I said to Sabit. "Maybe soon we'll get some heat."

Sabit just shook his head and clicked his tongue. But even he was intimidated by their gruff manner.

The computer company across the hall, run by local Koreans, had just modernized their offices. But they were already moving out. When I asked them about the sudden eviction, the boss just shrugged and said, "It's the KGB. What can you do?"

Downstairs Bozov was smiling.

"Robert! In Uzbekistan the *militsiya* are very strong." He explained that Mr. G had found new offices for the project, but they needed work before everyone could move in. We would stay until the *militsiya* threw us out. "But Robert, this is not *your* problem."

A few days later I moved our operations into our house at 45 Tazetdinova Street. Bozov was incensed. But I told him that the KGB had cut the electricity to our offices and it was impossible to work.

"But you will move into our new office very soon," he warned.

"We only have three months left on our contract," I reminded him happily. "And our house is very comfortable. There's no need for us to move again."

Bozov leaned back in his chair and eyed me slyly.

"Maybe you will stay longer than you think. You are just starting to understand the people of Central Asia, and to enjoy our beautiful women!"

Nadir shouted on cue, "Whoa, whoa!"

"You're just starting to become a *real* water expert!" added Bozov.

On a pleasant fall day Frank #1 came to lunch. JC Torrion had just returned and had never met Frank, who had completed his training with us but was staying on in Tashkent. Frank was sleeping in my old flat, in my old bed, with Shakhlo. The money she'd stolen from us was likely stored in her safe under the balcony, the one her mother had shooed me away from while she checked its contents. They'd been spotted riding around town, Shakhlo behind the wheel of the shiny purple Lada that she'd bought with pilfered money. According to reports, they both looked very happy.

Sabit arrived from the Alaski Market burdened with vegetables, fresh herbs, apples, pomegranates, a huge melon and a freshly slaughtered chicken. He took over the kitchen, preparing chicken soup and salads. As he generated dishes Ira and Mila set them out on the table on the patio among the bowls of plump cherries and juicy apricots they'd picked off the trees outside the front gate. Alisher, the owner's son, had ordered some local children into the trees and they were shaking the branches, bringing down the looser leaves. As they floated down, he raked them up. (Uzbeks are meticulous cleaners, especially where leaves are concerned.) He was also emptying the water in the swimming pool, draining it into the garden where the roses still bloomed inside the basil hedge.

"Mr. Robert, you want still to swim?" he asked obligingly. But his tone hinted that he thought October too late in the year for swimming.

"Why not!" I answered. The days were still sunny and mild and dunking myself repeatedly in the pool had become my morning wake-up ritual.

JC and Frank #1 smoked Gauloises and smiled and joked in the way of people made uncomfortable by circumstances. The subject of Shakhlo was being avoided, but her presence loomed larger than ever.

Akhror, the owner of the house, arrived midway through the meal. He took Sabit and me aside and said that moving our office into the house was illegal. The *mahallah* would either fine him or force him to kick us out. He was very upset.

Tashkent's *mahallahs* were not just neighbourhoods. They were local authorities that maintained social order. In traditional Uzbek society a revered elder, an *aksakal* or "white beard," who was consulted on important matters, led the *mahallah*. His role was like that of a tribal chief; he would sanction marriages or mete out discipline for violations. The Soviets commandeered the *mahallah* structures, turning them into local Soviets, committees that informed on residents. The *mahallahs* knew who attended mosques, who dressed in traditional costume, who was growing a beard, who was not a Communist. When the Uzbek government inherited them in 1991, they knew the *mahallahs* were too effective a tool to give up and continued to use them to intimidate and repress.

"Did Shakhlo talk to you?" I asked Akhror through Sabit. If anyone would know how to work the *mahallah* she would. Akhror quickly admitted that she'd been threatening him, saying it was illegal to accept U.S. dollars as rent on the house. If we didn't move our office out, she was going to inform the *mahallah* and the police.

"Blackmail!" shouted Sabit. "She's trying to blackmail him!" He clicked his tongue.

I felt the chain of command, from Shakhlo running up to Bozov and on to Mr. G. They'd never liked this house – we'd rented it without their approval – and now they were trying to force us out.

After some discussion, Sabit came up with a solution. He suggested we notarize a dummy *sum* lease agreement that Akhror could show the police and the *mahallah* if there was a problem.

"If she bothers you again," I said, "tell her to talk to me." I knew she'd never do that.

Relieved, Akhror opened up, letting loose a flood of spite: "Last summer when you were away Shakhlo ordered me to come here and turn on the sauna! Many different men! Many parties! My brother saw them too!" His brother lived across the street. Then in a whisper, "She's a prostitute!" Sabit translated all this with shock and delight.

Back at the table Frank #1 and JC were finally talking about Shakhlo. Frank was again offering to pay out the missing money.

"It's only to clear her name. Not because she's guilty of anything."

"If that's really what you want to do," said JC, a tiny smile appearing.

"I know that she never took the money from the safe." Frank turned to me. He exhaled his cigarette smoke and looked as defiant as Shakhlo often did. "You are completely wrong about her. She is only your scapegoat."

"Then who took the money, Frank?" I asked.

"I don't know. There were many people going in and out of that office all the time. Someone must have watched her open the safe and written down the numbers." But this sounded to me like Shakhlo's explanation.

"Really? And the key? Did it fall out of her pocket accidentally? Or maybe I was involved?"

He shrugged and lit another Gauloise.

A few days later JC and I were sitting at the long table in our living room, the big one where we'd hoped to hold our training workshops. So far we'd managed only two meetings here, Old Bozie having kiboshed almost every attempt I'd made to use the space. As JC answered e-mails, I was trying to come up with a work plan for our last three months. But I wasn't getting anywhere.

"Our training is a joke," I said with a hollow laugh. But JC didn't laugh. "Bozov thinks he won the Battle of Bishkek – I'm sure Nadir told him that he did. Sabit's report will soon change all that. But what then? We'll be back on the front lines again, fighting over something else. Whatever I propose, Bozov will shoot down. It's ridiculous. The game has gone way too far. It can't stop now."

JC lit a Gauloise and leaned back in his chair.

"What do we do?" I continued. "More training? But everyone knows the national teams are ineffectual. Even if they had the skills, even with their strategies in place, with real campaigns, they'd fail. These governments are dictatorships. They won't allow any real public participation. Anyway, the *will* to react is missing. People have too many other problems. Day-to-day survival kills any concern they might have about the Aral Sea and the coming water shortages. One day a really *big* water crisis is going to hit and create chaos. And nobody will be ready. They'll still be squabbling. And it will make the Aral Sea disaster look like a dried-up well."

I was ranting, being sensationalist, thinking out loud. *Letting off steam,* as Anatoly Krutov would say with a wink.

"The only chance we have now is to appeal to the rest of the world – build awareness in the West, get them involved. Get them to understand that the water crisis that may someday happen to you is already happening in Central Asia. There are lessons to be learned here. Pay attention! It's what Patrick Worms was supposed to work on in Brussels, but never got the chance.

"The other day Bozov surprised me. He handed me a pile of documents in English that summarized potential donor interest in the Aral Sea disaster. Most major countries were listed with brief comments. For example, next to Russia it said that blame for the disaster must never be put on them, that they had strong economic and military interests in the region, but little money. Some consultant had generated all this stuff and I'm sure Anatoly Krutov had passed it to Bozov and told him to give it to me. I can hear Anatoly pleading to Bozov: 'Kadirbek! Please! Make use of your foreign experts before they're gone. The Aral Sea project ends in 2003. Kadirbek, what then? Who will pay your salary?"

JC smiled. He thoroughly enjoyed our jokes about our squabbles with Bozov and Mr. G. But I also knew that he shared my concerns about the failure of our work.

"Forget the national teams and their campaigns. Come up with something big, something global. A PR strategy for the project. We'll take it to Bozie and see what happens." He smiled again and shrugged. "At this point, what have we got to lose?"

For the next couple of days Sabit and I put the Aral Sea disaster into a global context; we tried to connect what was happening here to the rest of the world. We searched the web, collected ideas, mulled things over. Then we went out on a long walk through the *mahallah,* ostensibly headed for a *gastronom* I liked, the one in "Little Paris" with the French food. As we strolled through the narrow lanes, I summarized what we had:

"It isn't just Central Asia that's running out of water, fresh water – the whole world is running out. It seems to be everywhere – in groundwater, in lakes and rivers and wetlands, in underground aquifers and springs, in glaciers, in clouds – but the supply is really fixed. Only three percent of the world's total supply is fresh and most of that is inaccessible. It lies in remote northern lakes or is frozen in the polar ice caps. Only one percent is actually available and we're losing some of that every day by polluting it.

"Since 1950 demand for fresh water has been growing exponentially, faster than population growth. In the United States – the world's leader in almost all resource consumption – the reckless use of water is depleting rivers, lakes and aquifers. Global warming is exacerbating the situation – the glaciers in the Rocky Mountains are disappearing and long-term drought is draining reservoirs. Cities like Atlanta and Los Angeles have already imposed conservation ordinances, restricting the watering of lawns and requiring that homes be fitted with water-efficient toilets and shower fixtures. But domestic water use is the minor culprit. It's in industry – in producing steel, in manufacturing goods, in generating power, in extracting oil from the ground – that water use is extensive. But by far the greatest use of fresh water is in agriculture. Irrigation sucks up 70 percent of the total fresh water supply in the U.S., and even in that advanced economy about half of this is wasted in the process. There, like here in Central Asia, the attitude to water has been greedy, sloppy and excessive. For a very long time everyone on the planet has been taking water for granted."

I was taking a lectury tone on purpose, talking to Sabit like a high school teacher. Despite his considerable knowledge and preference for the role of the *isham,* the learned instructor, he was happy enough to play the part of the impressionable schoolboy.

"But these days water is being reconsidered. It has become fashionable. And expensive. It's in vogue because it's healthy – doctors tell us to drink two litres a day. Bottled water, almost unheard of twenty-five years ago, is now a hugely profitable business. In Tashkent a litre of bottled water costs more than twice as much as a litre of gasoline. And governments all over the world are passing the responsibility for supplying and managing water to the private sector because they think that they can do it more efficiently. As supplies are being privatized, almost everyone is paying more for it."

"Terrible!" Sabit clicked his tongue. "Water is a gift from Allah!"

"It's also getting more dangerous. In many cities – Tashkent, Bishkek, Almaty, Moscow, St. Petersburg, Beijing – you can't drink the tap water. There's *E. coli, gardia,* all kinds of scary bacteria in it, and there are old lead pipes and polluted reservoirs. Visitors to those cities are warned not to brush their teeth in it. Even in Western cities like Toronto where fluorinated tap water has been drunk for decades, there are thousands of residents who won't touch the stuff anymore. As Mr. G so shrewdly pointed out in Canada there was Walkerton, which told us that even in water-rich civilized Canada, the supply is never 100 percent safe. And don't forget those terrorists walking around with tiny vials of poison, waiting for the order to hop the chain-link fence and pour it in our reservoirs. Pooosh! Four million dead. Just like that."

We stopped at a narrow canal, almost empty of water. We weren't sure where to go, all the streets seemed to be culs-de-sac, heading only for the canal. The houses were poor here, tin-roofed shanties, and the only faces we saw peeked around corners and glanced at us. They were mostly children. It was drizzling, but that didn't bother us at all.

"This way!" shouted Sabit happily. He led us back up the street we'd just come down.

"According to water experts, by 2025 world demand for water is expected to outstrip supply by 56 percent. Even allowing for a huge skeptical margin of error in this prediction, there's little doubt that water is going to get a lot more expensive. It will almost certainly be a commodity with a price that will be quoted daily, much as the price of oil and gold are today: Lake Superior Natural – $4.89 a litre; Greenland Glacial: $5.47 a litre. Countries such as Canada, Russia, Turkey and Finland will probably sell their water to the highest bidders. In many populated areas of the planet 'water-outs' will be as common as traffic jams are today. Many irrigated lands will have gone out of production and fruit and vegetables will cost a lot more than they do now. Home vegetable gardens will be popular. Innovative water technologies like drip irrigation, aquaculture, desalination, 'water-makers,' will be hot, and high-pressure aqueducts will crisscross the globe siphoning water to thirsty cities."

We were walking in circles. Sabit grinned, slightly embarrassed for not knowing his own city better. He took us up another street.

"Let's assume the worst," I said, knowing that Sabit shared my love of disaster scenarios. "The first global water crises will hit populous places where water resources are already being stretched to the limit: the Nile River basin, where water levels are falling. Beijing, where shortages have occurred. The Punjab and Bangladesh, where groundwater levels are dropping. The southwestern United States, where the giant Ogallala Aquifer is shrinking fast. Do you know that entrepreneurs who own the water rights in Texas are pumping up this water and selling it to the highest bidder? It's the new oil, and there's no law to stop them. And of course in Central Asia. In a not-too-distant drought year in one of these places, the reservoirs will drop dangerously, triggering strict conservation measures. Then will come the water-outs, the rationing of supplies and harsh punishments for those caught cheating. Then the first deaths, likely children and the old. Then the first UN water-relief tank trucks will roll into desperate areas of Cairo, Beijing or Dakha, opening their taps to allow thirsty residents a litre each. Or maybe it will happen in the inner city of L.A. Or just down the Silk Road in Samarkand."

"Robert, it's a terrible world! We are all so greedy. We don't appreciate our wonderful natural resources." Then he pointed. "Look! It's your favourite *gastronom.*" Sure enough, we'd come upon it almost by chance.

"Okay, here's the pitch: the Aral Sea basin offers the world a look at a water crisis in a nutshell. All the ingredients are right here: the overuse and depletion of limited water supplies, continuing drought, ineffective water management and unenforceable international agreements. Not to mention plummeting economies, social upheaval and political repression. *And* exploding population growth. The World Bank estimates that the five Central Asian states' population will double to 86 million by 2025. If water conservation measures fail – as so far they have – then within the next five years demand for water will overtake supply. Central Asia will be the site of the world's first major water crisis. And if the five states continue their bickering, the world's first water wars."

We were standing outside the entrance to the store now.

"'Whiskey is for drinkin', water is for fightin'!'" I said, quoting Mark Twain.

"Robert! Soon everyone will be fighting." Sabit was grinning madly, as if war would be a wonderful solution to the huge water mess.

"Central Asia's water crisis should catch the world's imagination," I said, walking into the store. I was keen on some Roquefort and a bottle of that good French Merlot I'd found here a few weeks before. "It will finally put Central Asia on the world map."

"Bozov will love it!" shouted Sabit behind me. "He'll have no choice. Robert, you'll save his job!"

Bozov's new office was a large room with a row of south-facing windows. The view was busy Navoi Street and the Chorsu Hotel, a three-sided concrete tower built in the Brezhnev era as Tashkent's leading hotel. It looked empty and decrepit. Near it was the Kukeldash Medressa, a minor copy of Samarkand's Registan. Beyond that was the huge Chorsu Market and old Tashkent, mazes of alleys snaking between packed mud-brick dwellings, mosques and *medressas,* some dating from the 15th and 16th centuries. Dusty and humble, each house had a courtyard garden that supplied the family with fruit and vegetables. But these old structures were not earthquake-proof and the city was gradually bulldozing and replacing them with uninspired concrete blocks festooned with cotton motifs. Old Tashkent was rapidly disappearing.

Sunlight poured through the rippled windowpanes, making the dust dance over the naked furniture that had yet to be properly arranged. Bozov and Nadir excitedly showed us where "Company BDPA" would set up. We got onethird of the room. The Uzbek national team, which had never had an office before, would set up in the centre – as a sort of buffer zone, I thought. Bozov and Nadir were stationed at the far end of the room, the map of Kyrgyzstan again on the wall over Bozov's desk. But they had a new addition, a middle-aged, red-haired Slav with a kindly face and submissive manner.

Bozov introduced us to his new secretary, Ludmila.

"As you see," said Bozov, "everything is prepared!"

"Things have changed," I said, smiling. But not everything. Nadir poured us tea. *"Limon?"* he snickered at me. "Oh! But I forget again to buy it at market!" He made his sour face.

Bozov was scowling at me in his authoritarian way.

"Robert! Your PR strategy is good." I had spun together our ideas, integrating them into a media kit, a website, a promo video and a press conference in Brussels, all designed to spark Western media and donor interest in the Aral Sea disaster. "We will begin it right away!"

Nadir nodded and looked pleased. *At last Ferguson did something useful!*

JC lit a Gauloise. He was ignoring his tea.

"Mr. Bozov, we will start on the new PR strategy as soon as it is officially approved and we settle this matter." He opened his briefcase and pulled out a piece of paper with BDPA letterhead. His nose twitched as he passed it to Bozov. "Here is a copy of the invoice I already sent you, for the work we've completed. Just in case you misplaced the old one."

Bozov glanced at the invoice and pushed it aside. He lit a cigarette, leaned back in his chair and grinned at JC.

"Monsieur Torrion, you are worrying too much about administration! Leave this to our secretaries." Ludmila looked up from her desk and smiled weakly. "Of course there will be no problem with this. But we will need a detailed financial report on all your expenses. Maybe there are some small charges that we will not be able to pay. Maybe some telephone calls your specialists made to Europe that we can't pay for."

He and Nadir smirked at each other. JC laughed sarcastically and shook his head.

"We must be very careful with our money," continued Bozov. He leaned forward and blew his smoke at JC. "We have a responsibility to the World Bank, to our donors." He was trying hard to look serious. "Mr. Guiniyatullin is very strict. He never pays for anything that isn't accounted for very carefully."

"Very, very strict!" echoed Nadir. "We must be *very, very* careful!"

JC got up. His nose twitched. He was infuriated.

"Mr. Bozov, I'm going to discuss this with Mr. Pearce at the World Bank. I'm not going to leave Tashkent until I have received assurance that it's going to be paid!"

Bozov eyed JC and sucked on his cigarette. He blew his smoke out at him again.

"Please give my regards to Mr. Pearce," he said with a sly grin.

The Abay Banya, like other bathhouses in Tashkent, was state-run, cheap, grotty and popular. It had become Sabit's and my favourite place to while away a weekend afternoon, cleansing the body, sweating away anxiety and bantering with strangers. On a Sunday in late October, after several games of badminton in light rain at the Walrus Club and a bracing swim in the icy Ankhor Canal, we introduced it to JC.

The old Uzbek attendant eyed JC and me coolly – foreigners were unusual here. He directed us to some old wooden lockers and presented us each with an ancient padlock – he retained the key – and a strip of thin cotton rag; devout Muslims cover their genitals. Tucked into our meagre wraps, we followed Sabit into the washing area, a round room finished in dingy ceramic tile with built-in benches and taps poking out of the wall. The domed ceiling was punched with holes and the afternoon drizzle dripped in. The taps leaked, the floor was slithery and steam from the showers billowed around the room. There was a smell of damp grunge and cheap soap. Blurry naked men lathered their bodies until they were white with foam and then dumped basins of water over their heads. Fathers scrubbed sons. Friends scoured each other's backs. A teenager washed down an old man stiff with arthritis. Waiting for their turn with a basin, a row of seated men watched JC and me with detached Uzbek smiles. We were minor amusements.

"Never wash before the sauna!" ordered Sabit. He gave us each a woollen hat, to keep the heat from escaping off the top of our heads. We donned them and entered the sauna. It had four cedar-planked platforms and we followed Sabit to the top where it was hottest. "A hundred degrees!" he claimed. In seconds our nostrils and ears were burning and it hurt to breathe. Sweat streamed off our bodies. Sabit ordered us down a level and took turns kneading our upper backs and shoulders. He chatted cheerfully to the other Uzbeks, ribbing the foreigners good-naturedly I guessed by the laughs he got.

"Robert, you have too much stress! You must see the masseur!"

After five minutes JC led us out – he couldn't take any more. Sabit ordered tea from the attendant and we sat in another round room next to the lockers; this one had no roof at all. The green tea and the cool drizzle quickly revitalized us.

The bathhouse was a series of contiguous circular cells, each topped with a cupola, like a honeycomb. It seemed to be styled after an ancient *banya* in someplace like Bukhara or Samarkand. But it was crumbling around us, from lack of money and indifference. In Canada or Europe such a place would be condemned, but here nobody seemed to mind.

The masseur was a stocky Tartar, built like a wrestler. He flexed his arms and pumped out his barrel chest like a bodybuilder in competition. His body was gnarled and disproportionate – his arms too big, his legs too short. He was very proud of it. He invited us to feel his biceps. He turned around and popped his shoulder blades in and out.

"Very impressive," said JC, laughing.

The masseur ordered me to lie down on a ceramic-tile altar. He yanked off my wet wrap and hosed me down with hot water. He bent back my arms and legs, cracked my neck and rotated my shoulders. He climbed on top of me, sat on my ass and twisted back my arms and shoulders. He kneaded my back, pinching and pummelling my skin until I nearly passed out. Finally he stood up and walked back and forth all over me. All the time he kept joking with Sabit who laughed his obligatory laugh.

JC passed when the masseur ordered him onto the altar next.

"He is sadistic!" he said, alarmed and amazed at the performance. Even Sabit declined. My body was numb, but back in the sauna it slowly unwound in wrenching aches, opening up like a piece of crumbled paper. I felt transformed, like I was a completely new configuration. We had a total of three rounds in the sauna. After we had rinsed off under a cold shower, Sabit led us into another round, cupola topped room with a pool. We plunged in. The water was icy – the same temperature as the canal – and we leaped around, laughing like schoolboys. Dressing in the locker room we looked at ourselves in the tarnished mirror. Our skin was pink and glowing.

"You see how youthful we are now?" said Sabit. "We are young men again!"

On October 27 David Pearce assured JC that BDPA's invoice would be paid. He asked JC to be patient. With this guarantee, JC flew back to Paris. But he was still furious.

"Since the meeting we had on the 23rd of October," JC wrote Mr. G on November 2, "no progress has been made. Mr. Bozov is delaying everything as he has been doing since the beginning of the contract. I have therefore decided to start reducing our expenses on site, and to concentrate on important work that has to be completed. As a first step, I asked Mr. Ferguson to suspend Mr. Tulenbai Kurbanov and Mr. Abbazbek Kasymbekov as from today . . . BDPA will suspend all activities until a guarantee of payment is received. I also asked Mr. Ferguson not to move to the new project offices."

17   Follow that Car!

"Mr. Robert," asked Alisher, frowning through the steady rain, "are you going to leave Uzbekistan?" Water dribbled off the rim of his blue baseball cap. In his hand was a hose that was spurting a limp stream across the sodden courtyard garden. I'd finally had to accept that the weather was too cold for swimming and he was draining the pool.

"January," I said, wondering if I'd really survive two more months of this.

"Because Shakhlo?" His expression was troubled.

I wanted to talk to him about the pointlessness of my situation – the games, the tricks, the lies, the expected payoffs, the blackmail. But young Alisher would probably already understand how things worked here. His father and uncle were connected to President Karimov's inner circle. They imported goods from overseas and sold them for dollars and made good profits. They were rich. In Uzbekistan you were either on the inside doing well or on the outside, poor and desperate. And as much as I liked Alisher, I couldn't see him bucking the system, or having any real empathy for me and my situation. Why should he? He had a great future mapped out for him: a university education, profitable businesses, a pretty wife, a big house in this *mahallah,* possibly even this house. After I'd told him everything he would probably ask, like Ira, Shakhlo and others, "Why didn't you just pay them off?" It was the way things worked here and you had to do it. *Not* doing it was the mistake. Everyone knew that. No, it was easier to blame Shakhlo than to try to explain my predicament to him.

"Shakhlo is part of the reason," I said. "It's complicated. Let's just say that our training is not going well. We're not yet saving the Aral Sea."

He gave me a sad but sympathetic look. Maybe I was wrong about him. Maybe he wanted to understand. Maybe his youthful perspective would shed some light on things. Maybe Uzbekistan was not as bad as it seemed. If there was any hope it had to be with the country's Alishers.

It was the first Sunday in November and it had been pouring all weekend. I was sitting out on the patio under the overhang, watching the rain that had already killed our erratic telephone line and the satellite TV. But after the months and months of drought, it felt strange. Remarkable.

The doorbell cooed deep in the house. Seconds later a short, athletic man in jeans, a black leather jacket and a captain's cap charged up to me across the courtyard.

"Sabit! You look just like a New York City cab driver."

"Robert!" he shouted. "You say such terrible things to me! And what are you doing? Anyone can walk in! The gate is wide open! You don't know how dangerous Tashkent has become! Every day there are burglaries and murders! People hacked to death for no reason at all! Tashkent is now *much* worse than New York!" He grinned as we shook hands.

"The drought is over, Sabit. In another week the Aral Sea will be filled and I can go back home. Mission accomplished."

"You did it!" he howled. "Congratulations!"

"It must have been that rain dance I performed a couple of nights ago. The next morning the clouds rolled in and the rain hasn't let up since. Sometimes the most complex problems have ridiculously simple solutions."

Sabit laughed again and then turned serious.

"They say there's flooding! But of course in Uzbekistan nothing can be proven. You never know if they're telling the truth." His eyes glistened. "Tea!" he shouted and disappeared into the kitchen.

Alisher propped a ladder against the house and climbed up onto the roof. He began toying with the satellite dish.

"The phone line is more important," I called up to him.

"Yes, Mr. Robert." He inspected the dish for a couple of minutes. "If you want, I give you private hospital phone line that we have at our house. Very good line. Very clear." He explained that someone had tapped into the line at a hospital two kilometres away and strung a link into the *mahallah*. "Line is free. No problem with officials. Only fifty dollars for new wire."

"Thanks, Alisher. But I don't think I'll be here long enough for that. I'll just use my mobile until the rain lets up. I can live without e-mail for a day or so."

Sabit rushed out with the tea and poured us each a bowl. We sipped silently and gazed at the saturated courtyard. Our plight now looked as bleak as the weather.

"Do you think this deluge is trying to mock our failed attempts at water conservation?" I asked.

"But Robert, we won the Battle of Bishkek!" Sabit's report on the regional workshop had scored a direct hit. But Bozie was rallying his troops and I suspected we were about to get massacred over BDPA's invoice.

"But the war's still on. Maybe we should take Boris's suggestion more seriously."

Boris Babaev, suspended along with Tulenbai and Abbazbek and anxious to get his withheld salary, had come up with a scheme to get both BDPA and himself paid. He'd suggested we lobby the national coordinators to support BDPA's cause. The national coordinators – the five governments' representatives whose job it was to approve all project work – were in their own battle with Guiniyatullin. Led by Tesha Avazov of Dushanbe, they had recently tried to take control of their national allocations for public awareness work. But in retaliation Mr. G, who used project money as his main lever of control, had withheld the salaries and expenses of all five national teams. As a result Component B was in limbo; nobody was getting paid or doing anything but fighting.

"Boris is Mr. G's spy," Sabit said. "You can't trust him. He's only thinking of himself." He shook his head and clicked his tongue. Then suddenly he brightened. "But of course he's right! Some of the national coordinators might support us. But only because they want to get rid of Guiniyatullin. Yes, Robert! You do have an opportunity!"

Sabit refilled our bowls and started strategizing. He put the national coordinators and the team leaders into two camps: those who might support us and those who wouldn't. His eyes were sparkling and he looked slightly mad. I let him talk, half-listening to him, but mostly hearing the pouring rain. It was drumming on the metal roof, and gushing off the spouts in the eaves troughs. It was knocking down the big yellow leaves off the quince tree and they lay flattened and drowned all over the patio. It was drenching the still-green basil hedge and thrashing the grapevine that Alisher had pruned back. It was smacking the paving stones and surging in torrents under the steel gate and into spreading pools on Tazetdinova Street.

By the time the tea was finished, I had agreed to lobby the national coordinators. Sabit was overjoyed. But I didn't share his optimism. I didn't trust the national coordinators. I didn't trust anyone any more except Sabit. I was feeling fatalistic, convinced that foreigners just don't win in Central Asia. Not through lobbying and diplomacy anyway. The only way to win was to fight a war.

One morning a few days later the doorbell cooed impatiently. I opened the gate and faced two militiamen holding up Sabit by the armpits. He was lopsided and twisted, like a rag doll. He looked short, angry and scared.

The *militsiya* were accusing Sabit of being a Wahhabi, a member of the Muslim fundamentalist sect that President Islam Karimov blamed for the 1999 terrorist attacks in Tashkent. Apparently his beard was the problem. Devout Muslims often wore beards – including Wahhabis – and in paranoid Uzbekistan, the only beard that wasn't suspect was the wispy white beard of a revered elder, an *aksakal.* To me, Sabit's trim grey-flecked fringe along his jawline – he shaved his moustache – made him look like an Amish farmer.

The policemen had never heard of the Aral Sea project, IFAS or the World Bank. I produced documents with Sabit's name on them and gave them my card. Finally they shook my hand curtly, got back in their car and drove away.

Sabit gradually calmed down over a cup of tea.

"I think Mr. G's trying to frighten us," I said, wondering if this harassment was the start a new campaign to get Sabit, like the one he ran successfully against Victor. "I think he's punishing us for winning the Battle of Bishkek."

"It's a warning," Ira said, sitting at her computer. "To stop asking the national coordinators for support."

Sabit and I looked gloomily at each other. He had called most of them and got their verbal agreement to help our cause. But none of them had sent a letter of support.

"Central Asians will never go against their own people to support a foreigner," said Ira forthrightly.

Sabit scowled and clicked his tongue. Ira ignored him.

"She's on *their* side," he said later at lunch. "She and Boris. They're both against us. They're *traitors!*"

"After several weeks of letters we still have not found a way out of this current impasse," wrote JC to Mr. G on November 22. "In Mr. Bozov's last letter of November 17, he asked BDPA to provide a final report. I would be grateful if you could send me and the World Bank an official letter requesting the project to be stopped if you really want to stop it. BDPA never intended to stop the project nor have we had any discussion on this with Mr. Bozov. We are preparing for the World Bank Mission in early December, and are ready to assist the national teams to finalize their strategies."

José Bassat had e-mailed me from Washington with news of the World Bank's full review of the Aral Sea project. As part of it, he asked me to submit a confidential report on our work.

"It's our last chance," I said to Sabit.

Sabirjan straddled lanes and zoomed around cars on Sharaf Rashidov Street. I was in the front seat, my eyes glued to the traffic. Sabit, in the back, spurred him on in Uzbek, making it clear that this was an emergency. Sabirjan, taking on this task with the coolness of a Formula One racer, apparently had no qualms about speeding through the city.

We were headed towards the World Bank offices on Academician Suleymanov Street. Three times that morning I'd tried to reach Anatoly Krutov and twice I'd been patched through to a busy signal. On the third attempt I'd asked the receptionist if he was telling Anatoly that it was me. "Of course," the receptionist had said. "I recognized your voice." Then I'd said to Sabit, "Okay, if he won't talk to me, then let's go get him!" Sabit, of course, was ecstatic.

A week before I'd sent my confidential report for the World Bank's review of the project to José Bassat in Washington. I'd also provided a copy to Anatoly, who was coordinating the review team. In it I'd said that the director of Component B and the head of all components of the Aral Sea project did not want this contract from the beginning, that there had never been any attempt on their side to cooperate with BDPA. I'd identified some of the practices within Component B that were leading to its failure. Sabit had helped me hone the language until we were satisfied that every point was clear and our case solid.

A few days later José had e-mailed back, to thank me for my report and to tell me this: "Anatoly has confirmed the news I got this morning that due to budget cuts I have to skip this mission. Not the wisest of decisions taking into account the critical moment for this component . . . someone probably with little communications experience will look into the component and I will be commenting via e-mail/phone."

Later that day Nadir had called, asking for Ira's help in translating a report for the World Bank. A few minutes later our fax machine had begun spewing out pages of my confidential report.

"Nadir, how did you get this report?" I'd demanded.

Nadir had squirmed. "From Mr. Guiniyatullin. He asked me to make translation. But it's too difficult." Then in a simpering voice, "Can Ira make it for us?"

"That idiot Nadir!" I'd shouted around the office. "He doesn't even know what he's got his hands on!"

Sabit's eyes had nearly exploded. "It's Krutov!" he'd yelled. "Anatoly Krutov gave it to Guiniyatullin! It's the only way. I told you not to trust him!"

That was on Friday. Anatoly was in Almaty with the review mission. He would be back the following Monday. It was now Monday.

We rounded the corner and pulled into the entrance of the World Bank compound just as Anatoly, in the back seat of a sleek black Toyota Cressida, sped off in the opposite direction.

"Follow that black Toyota!" I hollered. Sabit shouted the Uzbek version. I grinned at the World Bank security officer standing at his guard post. He glared back as Sabirjan spun gravel before we sped away.

On Navoi Street we caught up to the Toyota at a red light. Diminutive Anatoly, sitting primly dressed in a dark suit, looked like a rich schoolboy being taxied to prep school. Sabirjan tapped his horn. Anatoly's driver, a big Slav, frowned at us with dour self-importance. (Drivers in Central Asia can be powerful behind-the-scenes players. Soheil Ramanian, our go-between, had told me that the head driver of the Aral Sea project was Mr. G's inside contact. And World Bank drivers, I figured, would be very high on the driver totem pole.)

Anatoly glanced at me but quickly turned away. The light turned green and his car charged ahead. Sabit ordered Sabirjan after it and we raced down Novoi Street, weaving in and out of lanes, trying to catch the black sedan. Sabirjan's eyes were fixed straight ahead, intent on winning. I'd never seen him like this before. Our chase was bringing out a repressed wild side. I had a new respect for him.

"He's headed for the project office!" I yelled as we bounced through an intersection in front of the flying-saucer state circus. The light behind us popped from yellow into red and I turned around and watched the Toyota screech to a halt. "We got him!"

Five minutes later Sabit and I were standing in the project parking lot as Anatoly pulled in. He got out of the black Toyota and greeted me with a cheerful smile.

"Robert! So good to see you! I'm so sorry, I've been very busy with the mission that I haven't had time to speak with you."

"No, but you've had time to pass my confidential report on to Mr. Guiniyatullin."

Anatoly dropped the smile and gave me a look of consternation, as if this was indeed very disturbing news. But then he brightened, reviving his old sunny mask.

"You know this is the best way," he confided. "You see, it's a ver good report. In fact, it's excellent! Mr. Guiniyatullin now knows about everything that's been going on. *Everything!* Now he'll have no choice. He will have to pay BDPA."

"Anatoly, you had absolutely no right to show that report to Mr. G. I trusted you."

Anatoly looked regretful. Then he turned gleeful again.

"But you don't understand! It's already worked. Mr. Guiniyatullin has already agreed. BDPA is going to be paid! Please give the good news to Jean-Charles and send him my best regards." He started to move away and then added, "And Robert, when this review is over, we will talk about extending your contract." He winked. "Now I must run! I have a meeting with Mr. Guiniyatullin and the review mission."

"Anatoly! When am I going to meet with the review mission?"

But he disappeared inside the building.

Sabit shook his head and gaped at me. I watched his eyes grow larger and larger, moister and moister, taking on the craziness that these days seemed to possess him more and more.

"Robert!" he erupted. "He's such a liar! He's the biggest liar in a country that is full of the biggest and best liars in the world!"

His voice bounced off the walls of the Kukeldash Medressa, then off the five-storey concrete building housing the Aral Sea project. It echoed around us in the blacktopped parking lot. We were alone, standing amid a dozen parked cars.

"Liar! Double-crosser! Traitor!"

The traffic on Navoi roared by, oblivious to Sabit's shrieking.

Then he was howling with laughter. And I couldn't help it, I started howling too.

Maybe he really was the devil: the sunken bloodshot eyes, the gigantic bald pate, the preternatural pasty skin, the flabby jowls. The pinched mouth that barely opened when he spoke. The tiny yellow teeth. The Marlboro addiction – of course the devil would smoke; smoking is evil. Maybe his rasping growl was the devil's voice. Maybe his piercing glower was the devil's stare, the devil's way of penetrating you, boring through your eyes into your soft pliable mind to flesh out anything that contradicted him.

Or maybe he was just a bitter old man, caught in a purgatory between East and West, between communism and capitalism, between the old ways and the new.

Mr. G's new office was bigger and sunnier than his old one. It was more spacious too and felt a bit empty, as if more furniture was needed. But reassuringly the same map of the Aral Sea basin was covering most of one wall, and the Aral Sea was still vast and baby blue. Some things never change.

It was Friday afternoon, December 15. Ira and I were sitting opposite him at a small table that abutted his desk. All week the rumours had been flying: BDPA is paid! BDPA is not paid! Our suspended local specialists were sure that we were: "So now you can pay me my salary!" demanded Boris Babaev. "Mr. Bozov says it's true!" claimed Tulenbai. But Ira was sure that we had not been paid: "He will only pay you when you agree to pay him." And Sabit kept looking at me, shaking his head and clicking his tongue. "Blackguard! Double-crosser! Two-timer! Liar! Double agent!" I wasn't sure if he meant Anatoly, Bozov or Mr. G. Or all three.

Finally I decided to just ask the monster myself.

Rim Guiniyatullin took off his dark glasses and lit a Marlboro. He told me I looked good. Relaxed. I must be resting well in my luxurious house. Then he smirked, turning the remark into a sarcastic dig at our big house, which he had never "approved."

I avoided his stare by gazing over at his baby-blue Aral Sea. What if I had done what he and Bozie wanted? I decided to start all over again. I zoomed back eight months and began reliving my contract the way my bosses would have had it:

First off I fired Victor. I did it as soon as Bozov enlightened me to the reality that a "Korean" could never understand the complex people of Central Asia. Then I hired all Bozie's local specialists. I soon had to sack Sabit when I realized what a troublemaker he really was. Luckily Bozov had a replacement waiting in the wings – the wild-eyed Kazakh with the vodka breath I'd met at the start of the project. He, Bozov and I became good buddies and spent Friday nights at the *chaikhana* on Babur Park, drinking and flapping with a fast crowd with excellent *mafiya* connections. I rented Mr. G's huge flat and took five others for our consultants, all way overpriced, but what the hell, it made him happy. After Bozov's gentle prodding, I persuaded JC to cover tuition for Mr. G's two sons in the U.S., hinting that a contract extension for us was in the works. More prodding from Bozie got Valentina and me out on a few dates, but we soon agreed to just be friends. I adopted her strategy as the model for the region, laminated five copies of her Escher-like flow chart and the team leaders put them up in their offices. It was truly a work of art and they studied them scrupulously whenever they ran into a public awareness dilemma. Agreeing we needed a more positive approach to the Aral Sea disaster – enough with the bad news already! – we focused our campaigns on upbeat messages that dispelled fears that water was going to run out. We created the Aral Sea Lottery (Bozov's idea), a chance for farmers, water managers and government officials – and their friends – to win a weekly pot of $1,000. We broadcast the winners every Saturday night, using the event to air our new public awareness spots, excerpts from schoolboy essays on reviving the traditional respect for water: "No spitting in the canals!" And as the Prophet Mohammed says, "Eat, drink, but do not waste!" I ignored Shakhlo's indiscretions, raised her salary and made good use of her connections. She and Frank #1's wedding was the highlight of the third regional workshop, held in Osh during an ethnic clash (we were pretty lucky, we only lost four participants). And now I was sitting here with my good friend Rim, sharing a bottle of Canadian Club (Rim had claimed it was better than vodka, but I refused to agree) and discussing the extension of our contract. I was suggesting six months, but Rim was insisting on at least a year.

My eyes drifted back to Mr. G's glower and I plunged back to Earth. He was talking about the review of the project. There were some small problems.

"The World Bank is asking us to change some of the project's objectives. They don't realize how difficult this is. They can only see things from their own point of view." He bared his little yellow teeth. "Maybe tomorrow I will know about Component B. But everyone agrees this component is working well. There are no problems here."

I asked him why I hadn't been allowed to report to the mission.

"Mr. Torrion suspended your operations on November 2. He stopped your contract. Your work here is finished."

He knocked the ash from his cigarette into the onyx ashtray. Then he picked up a document and slid it in front of me. It was my confidential report for the World Bank.

"When I started reading this, I knew I was reading something I shouldn't have been. Sometimes what is written on paper can be very dangerous." He dragged on his cigarette and stared at me. "You blamed me and Mr. Bozov for certain things. You said it was my fault that one of your international consultants did not come. But these things are not my fault. And they are not Mr. Bozov's fault. They are Mr. Torrion's fault. He refused to cooperate with us. He didn't follow the terms of reference in your contract."

"What about BDPA's invoice?" I asked.

He looked away and said nothing for a minute.

"I have to answer to the donors for work done, or not done. We must agree on what BDPA work to pay for. What we need is a gentleman's agreement."

"What's a gentleman's agreement?"

He bared his yellow teeth again, amused that I didn't know his terminology.

"It's a private understanding between gentlemen. There is no need to sign any papers. We can discuss things and reach an amicable understanding. It will not be difficult. You can pass this message on to Mr. Torrion." He looked bored for a few seconds. Then he butted his cigarette and leaned closer. "If you ask for arbitration you will lose," he growled. "If you try to discredit me or the Aral Sea project I will hire a lawyer. It will cost only $200. And I will find you, wherever you are." He paused for a few seconds, then added, "I will use the internet to find you!"

He glared at me for a long time to make clear his threat.

I glanced at Ira who was staring uncomfortably at the table. Then I looked around his office. He had no computer. He probably hadn't a clue how the internet worked. And only $200 for a lawyer? *Mr. G, you are so cheap! Am I not worth more than that?*

Mr. G put on his dark glasses and rose from his chair. But he stayed behind his desk; he didn't want to shake my hand. Ira and I got up and walked towards the door. As we neared it, he growled again, but now it was a worn-out growl, more a low grumble: "The Aral Sea project's reputation is shit. Everyone knows that."

I turned around. He looked like a sad old monster that had grown bored by his own tricks. But I could see he was speaking the truth. He knew his project was shit.

"So what does it matter?" he asked. "You can say whatever you want."

18   Suspects and Alibis

The day after my meeting with Mr. G, Shakhlo was hacked to death in her bathtub.

Sabit and I learned about the murder two days later, on Monday morning, December 18. We were sitting on the couch watching BBC World on TV – vote recounting was underway in Palm Beach, Florida, to see if it would be Gore or Bush. We were in gloomy moods. The flimsiness of American democracy seemed to be underscoring our own reckless state of affairs. The whole world seemed very wobbly. Realizing that Anatoly and Mr. G had shut us out of the World Bank's review, we felt defeated. There was nothing left to do. We'd lost our last chance. BDPA was going to get shafted and JC would have to submit to Mr. G's "gentleman's agreement."

Ira arrived and quietly began work on a translation. After a few minutes she started talking in a sombre voice:

"Shakhlo was found dead in her flat on Saturday morning. The police said she was stabbed 26 times. They thought it was two people who did it, a man and a woman. Nothing in her flat was stolen."

Sabit and I gaped at each other. His expression was a mix of shock and delight.

"She did something *really* terrible!" he shouted. "She finally went too far!"

Ira was staring blankly at her screen. She had spoken methodically as if she were reading an item on the evening news. Now she ignored Sabit and turned to me. I could see how upset she really was.

"Robert, we are all being accused. The police are calling in every member of our team."

I was remembering that Sabit and I had just been talking about Shakhlo the other day. After one of our lunches at his favourite *chaikhana* on Navoi Street, we'd walked past a new hair salon called "Shakhlo." "It looks like Shakhlo has got herself a new business," I'd joked. But Sabit had only clicked his tongue and hissed, *"Horrible woman!"*

The phone rang.

"Good morning, Robert." Tulenbai's voice was lucid; he wasn't drunk, not this time. "I come now from police. They show me pictures.

Shakhlo in bathtub. Her stomach cut open, her guts falling out everywhere. Everywhere blood. Bathtub is full of blood. Her eyes wide open. Robert, is very terrible."

"Yes it is, Tulenbai. What did the police ask you?"

"They ask me if you and Sabit and Victor steal money with her. Then kill her."

"And what did you say?"

"I said no, Robert is good man. He is my friend."

"Thank you Tulenbai. And Sabit and Victor?"

"I said Sabit don't like her, but not kill her. But they want Victor. They think Victor is real murderer."

"That's ridiculous, Tulenbai."

"Detective Karimov will call you. He want to meet you in his office."

"I'm sure he does. Thank you, Tulenbai."

Sabirjan zoomed us down Ghafur Ghulom, a wide avenue lined with naked poplar trees and identical concrete apartment blocks. Ever since our car chase a week before he had been in Formula One mode. Everywhere we went now we raced, as if he was discreetly reminding us that our time was quickly running out.

It was the half-dead afternoon of one of the shortest days of the year. The temperature was just above freezing and the Central Asian sun had been banished behind a heavy blanket of cloud, as far away as possible. We passed a new state bank, glistening chrome and black, as garish as a Cadillac, then skirted the flying-saucer state circus and ducked into a concrete conduit. We came up onto Furqat Street, passed a row of tacky furniture stores and veered onto Halqlar Dustligi Street at the monstrous Palace of the Friendship of the Peoples. I had come here shortly after I'd arrived – with the officer who'd shown me the flats I was expected to rent but didn't – to see the monument to Alisher Navoi in the dried-out park behind the palace. Navoi was popularly called "Uzbekistan's Shakespeare." We flew past the new-fashioned white marble and tinted glass Oliy Majlis parliament where Uzbek deputies rubberstamped the president's decrees and patted each other on the back.

Stopping at an intersection, I zeroed in on a shop advertising Korean and Chinese appliances, swapping glances with a sharp-eyed clerk leaning on the counter. He waved at me and grinned and I recognized him as a young Uzbek who'd once stopped me on the street to offer his services as a tour guide. After a friendly chat, he'd hit me up for a $2,000 loan so he could buy a flat. "It's not so much money for you," he'd said. Then it came up that he knew Shakhlo – she had so many "friends" – and I'd seen that he shared her audacity.

Images of her were flooding my head. I saw her again at the *chaikhana* in Babur Park, at the start of my contract. She was smiling at me in her beguiling way, beckoning me to join her in an arm-fluttering dance. Then we were flapping and shaking and jiggling while Aral Sea project staff laughed and cheered us on. Then I remembered her washing her new purple Lada at the house that hot weekend in August just after the return of the Franks. She'd giggled like a schoolgirl and her nipples and her belly dancer's belly had shown through her wet shirt; the same belly that was now slashed and gutted. Later she and Frank #1 had stuffed *plov* into their mouths with their hands and laughed. And Shakhlo had said sarcastically, "We should all be very thankful to our team leader for arranging such a nice party."

Sabirjan pulled up in front of the Chilanzar microdistrict police station, a five-storey hospital-green building set back behind a high steel fence branded in cotton motifs. Security officers let us through and Ira and I walked up a dimly lit stairwell to the fourth floor. She knocked on a door and it opened into a cheerless room jammed with silent Uzbek men, all uniformly hunched forward on rickety chairs. They stared up at us with hollow eyes. I recognized Faizillo, Shakhlo's driver, and he acknowledged me with a slight nod, rather amicably I thought under the circumstances. Beside him was a bored-looking man with a moustache who I later learned was Shakhlo's ex-husband. The detective behind the desk ordered us down the hall to another office. But Detective Karimov was out, so we waited. Ira was nervous and withdrawn. We gazed out the hall window and said nothing.

Detective Numan Karimov interrogated me coolly, smiling as he provoked me with questions full of insinuations. He wanted me to think that he didn't believe my answers, that I was a prime suspect in Shakhlo's murder. During the interrogation I found myself avoiding his furtive smile by gazing beyond him at his bright colour posters of SAMARKAND, BUKHARA and KHIVA, just as I'd preferred the huge map of the Aral Sea to Mr. G. I tried to identify the turquoise and gold cupolas and minarets in the pictures, amusing myself with the idea that, having visited all three sights, I was now some sort of expert on Central Asian architecture. *Of course, that one's the Abdul Aziz Khan Medressa, the one with the garden of poppies out in front.*

"Shakhlo Abdullayeva was alone in the flat on Saturday morning," said the deputy chief of homicide. He was piecing together the crime scene. "Faizillo, her driver, had taken her two children away so she could complete some business. She was selling the flat. You see, in a few days she was going to go with her two children to Brussels to live with your foreign specialist." He glanced at the open file on his desk and after a few seconds added, "Frank Thevissen." He smiled slightly and waited to see if I would respond.

I had had no idea she was going to Brussels to live with Frank. But I said nothing.

"Faizillo called her just before eleven, to ask her if he should bring the children home. She told him not yet. We think that two people were with her, that she was discussing some business with them. When Faizillo called again about an hour later, there was no answer. He arrived at the flat with the children shortly after twelve o'clock and found Shakhlo in the bathtub. She had been stabbed a total of 26 times, more than enough to kill her." He paused for effect, then added, "The children saw her as well." He was watching me closely; the little smile seemed fixed. "We believe the perpetrator was a woman. A man probably assisted."

"Why do you think that?" I asked.

"Because of the nature of the stab wounds. The blows were lighter, more like a woman's. And because of the number of blows, we think it was a crime of passion." He paused again. "Love. Deceit. Jealousy. Murder."

I wondered if you could really distinguish the sex of the killer by the vigour of the blows. But then I had no reason to think Karimov incompetent. In fact his casual dress and cocky manner suggested that he enjoyed his work, that he was good at it.

"Maybe the perpetrator was a small man," I said, egging him on a little. "A teenager or a slightly built male with a light touch."

I was making a dig a Faizillo. He was skinny, but I knew his blows would be anything but lady-like. In the room down the hall he'd looked unruffled, not at all guilty, more like a patient waiting to see his doctor for a minor checkup than a murder suspect under detention. But I strongly suspected him. I recalled the day Shakhlo had massaged the back of his leg in my flat. "My nephew," she'd called him with that slightly mocking tone. But Faizillo looked Tajik and not at all like Shakhlo and we all knew they were carrying on together. (Shakhlo the chronic liar; she disliked telling the truth, even when the lie wasn't really necessary.) Surely here was Karimov's crime of passion. Forsaken Faizillo, jealous of Frank #1, had butchered his lover.

"Is Faizillo a suspect?"

"We have detained him," said Karimov. "But he has an alibi." Still the smile. "He was with her children all morning. Besides, he's cooperating, giving us leads."

Cooperating. Such a loaded word in Central Asia.

"Where is Victor Tsoy?" he asked. He was looking completely serious now.

"I don't know. I haven't seen him in several weeks."

Then he drilled me about Victor: Why had I fired him? How much did he steal from the project? Were we good friends? Were we still working together? I said that the project bosses had demanded I end his employment, but that he'd done nothing wrong. No, we weren't working together, but yes, we were friends.

"Victor Tsoy is not your murderer," I said.

He raised his eyebrows to show that again he didn't believe me.

I looked again at his posters. *That would be the Mohammed Rakhim Khan Medressa in Khiva. It faces the Ark. It was named for Khiva's last Khan, the one that was also a poet.*

Then Karimov dismissed me. He said he had some questions for Ira alone.

I went out into the hall and looked out the window. The late afternoon sky was turning dusky. Children were shouting and laughing below in a playground wedged between identical apartment blocks. Trolleys, Volgas and Nexias whooshed up and down the wide avenue. The familiar sounds of a Tashkent afternoon.

"I'll tell you outside," said Ira thirty minutes later. We walked down the dark stairwell, out the building and through security. When we got in the car she asked Sabirjan to take her home. Then she stared out her window for a minute. We sped back along Halqlar Dustligi Street, zooming through the traffic.

"He has a theory," she said finally. "He thinks that you and Victor and Sabit were involved in the murder, that you were operating together. He thinks that you were stealing money with Shakhlo. And that all of you were sleeping with her."

I laughed. She gave me a disapproving glance.

"I told him that it was impossible," she continued, "that the three of you didn't even like Shakhlo. Then he asked me if you and Sabit and Victor were blue."

I laughed again. "Blue" meant homosexual. I'd heard that gays were frequent targets in criminal investigations in Tashkent. Homosexuality was illegal in Uzbekistan.

"But he already said the murderer was a woman!"

Ira looked out her window again. She didn't want to talk about it any more.

Sabirjan stopped the car in front of her block. She opened her door and turned to me.

"It's Victor they really want. They're looking for him everywhere. We have to warn him."

"You'll be honoured to know that we're prime suspects – you, Victor and me. Detective Karimov has an inspired theory that we were operating as a sort of blue triad, stealing money from the project with Shakhlo. Even though he thinks we are gay, we were also having sex with her. Then we killed her."

Sabit hooted. "Tulenbai has been telling them such stupid things! But Robert, it's good! Because it's too crazy. No one will believe these stories."

We were sitting in the living room of the house on Tazetdinova Street, drinking tea as usual. It had been a long day. We'd begun it feeling despondent. But now we were revived. We had something exciting to immerse ourselves in. We had Shakhlo's murder.

"I wonder what *really* happened."

"Of course it was *mafiya*!" Sabit shouted. "She tried to cheat someone and he got her." He shook his head and clicked his tongue again. "Robert! She had so many enemies. But will Karimov ever find the right one?" He grinned.

"So far it looks like he hasn't got many facts to go on. That means he's still going to go after us – you and me and Victor. Please, Sabit, you have to be careful."

For a Wednesday night the Salty Dog was dead. The only other customers in Tashkent's most popular expat hangout were two Brits standing near the big-screen TV, swapping loud opinions on the rugby match. I wondered if everyone had found somewhere else to go. But probably it was just Christmas. The foreigners were all escaping.

"Robert, how is your crazy project?"

190

Victor sat down opposite me on a bank of upholstered seats. His relaxed smile suggested he knew nothing about Shakhlo. He had finally called just an hour before and I'd suggested we meet here.

"Crazier than ever. I guess you haven't heard the news."

He took a big swig of his draft Holsten and puckered his moustache, now coated in suds. He was Samarkand Victor, at ease and sardonic.

It had now been four days since the murder and 24 hours since my meeting with Karimov. The detective had been calling, asking to see me again. But Ira had lied each time and said I was out of the office. Anxious to get out of town, I'd changed my flight. Very early the next morning I was flying to London for Christmas.

"I just came back from an NGO conference in Bukhara," said Victor. "I love Bukhara. Very quiet, nothing ever happens there. Did I miss something exciting here?"

"Were you there on Saturday morning?" I asked. He nodded and gulped more beer. "Then you have an alibi. You should be fine." He gave me a puzzled look.

"Robert, Bozov has taught you well how to play games."

"Shakhlo was stabbed to death Saturday morning. The police are investigating every member of our team. At the moment you're their prime suspect."

His solemn eyes froze on me and I watched Samarkand Victor dissolve into brooding Victor.

"I don't want to think about that disgusting woman."

Over the next few minutes I told him all I knew. As I talked he watched the TV. He didn't make any comments or show any emotion.

"She got what she deserved," he said coldly when I'd finished.

"I wouldn't say that to the police."

He shrugged and looked indifferent. "The police can check my hotel in Bukhara." He stroked his moustache with his forefinger, wiping away the beer suds. "I told you before, she was professional KGB."

"KGB? Well professional something, anyway," I said with a smile.

Victor took another long drink and puckered his foamy moustache at me.

"There are many people who could have killed her. She was a greedy dishonest woman. She had many enemies." He looked back at the TV and after half a minute said, "It costs $2,000 to have someone killed in Tashkent." He said it matter-of-factly, like something you'd quote from a tour guide. I wondered how he knew the price but didn't ask. "Anyway it doesn't matter who killed her."

"Victor, you were their scapegoat before. Mr. G might be setting you up again."

He frowned arrogantly. "The police have nothing on me," he said.

We both gazed at the TV. The game was over and the two Brits were leaning on the bar, gaping at Britney Spears leaping about in her schoolgirl uniform as she sang, "Hit me baby one more time!"

"Don't worry, Robert," he said, turning back to me. He smiled warmly. "Thank you for telling me this. Now I know what I will do."

I was packed by midnight. Only six hours to departure time. At 2:00 a.m. Sabirjan would take me to the airport. I sat down in front of the TV and my mobile phone chirped.

"Robert!" said Sabit loudly. Too loudly. "They're taking me to the station!" There was panic in his voice. "They are very rude, standing here telling me that I must go. I said that I don't have to go, but now they are forcing me. I told them I was working for BDPA, for a French company with an excellent reputation. On the Aral Sea project for the World Bank. I said that I was a Sufi and a poet! I showed them my book. But they don't understand anything. They're complete idiots!"

In the background I could hear his wife crying.

"Sabit, calm down. Remember your alibi. You were with me Saturday morning. We worked all morning on those awful receipts for Bozie. Then we went to the Ankhor Canal for a game of badminton. As usual you won. We talked to the old men playing backgammon. Then a brisk swim in the canal. Later we were at the Abay Banya . . . "

He interrupted me with a desperate laugh and then the line went dead.

192

19 Lies, Lies, Lies

Armorer Wason shivered and sipped her pint of bitter. "You're not going back there," she said firmly. We were sitting on a crowded patio next to the white pavilion in the middle of Hampstead Heath. It was a snowy afternoon three days before New Year's and the English holiday spirit was in full force. People trotted by on the walkway with their dogs and prams and scooters and talked in cheerful animated voices.

Ever since our projects had overlapped in Mongolia, Armorer (her background was Welsh) and I had made a pact to share a good meal whenever we found ourselves in the same city. So far we'd had surprisingly delicate veal scaloppine in a freezing Italian café in Ulaanbaator, exceptional Sichuan-style chicken with peanut sauce in a hole in the wall in Beijing and most recently Georgian *khachapuri* and *khinkali* at the Gruziya restaurant in Tashkent. Now it would be London. We'd just spent the past hour tramping through Hampstead Heath's fanciful meadows. Armorer, in her wellingtons, had led me like a trooper through fields of mucky grass and half-melted snow that was stuck to the unruly hedgerows and monster oaks. All the while I'd been filling her in on the latest news from my feckless public awareness project.

"You'd be crazy to go back there." She frowned as she looked out on the calm wintry scene, trying to cleanse her mind of the messy images I'd just conjured up for her: Shakhlo's bludgeoned body and an array of dubious suspects, all grasping bloodied knives and wearing vengeful grins. Armorer's philosophy matched our walk in the park: keep your boots clean, and if you find yourself in the muck, walk briskly onto drier ground. She eyed me sensibly. "It's not your problem. Get your money and get out."

Here in the middle of a protected patch of English countryside in North London it all was sounding far away and ridiculous. Farcical. It was a tragicomic musical straight out of Bollywood. A comedy duo with a bad shtick. A bald monster that growls "I'll get them!" A selfdeprecating vodka terrorist. A matronly half-deaf Communist moved to tears over a bouquet of roses. A dictator who gives every one of his subjects a watch with his squirrelly face on it. A sunny Russian midget who can't stop lying. Two jolly fellows named Frank. A smiling police detective with an office like a travel agency. An empty safe. And a bewitching belly dancer with a shiny new purple Lada who ends up gutted in a bathtub. Everyone fluttering, flapping and wailing across the screen. And in the background, a baby-blue sea that keeps retreating.

"You're right of course, I shouldn't go back. But I should have gotten out months ago. In fact, I never should have taken this job. It was loaded from the beginning: All those months it took BDPA to negotiate the terms. The gaping holes in our contract. Bozov's dense belligerence and Mr. G's wily tricks. The campaign against Victor. The sabotaged workshops. The World Bank's bogus support." I grinned. "But we had our little successes too."

Armorer eyed me skeptically.

"I think I knew all along we'd never win. But sometimes you just can't stop, even if you know you should. You plummet off a cliff thinking that some ledge will be there and you'll land on it and you'll be battered and bruised but you'll bounce back up again. Like in the Bugs Bunny and Road Runner cartoons. You can't die. But when Shakhlo was murdered I suddenly realized how far things had gone. I'd crashed on the bottom. There was no ledge to rescue me."

She was frowning at me, looking a little worried, as if I might have a gambling habit that needed professional attention. Her thin rosy cheeks matched her bulky red ski jacket. She'd twirled a canary yellow cashmere scarf around her neck and pulled a blue woolly hat jauntily over her head.

"Well, you certainly weren't bored!" she said with gusto. "I think that's why you really stayed – for all the entertainment!"

"It's not over yet. The World Bank will have completed their review when I get back. Maybe Mr. G's house of cards will collapse and the corruption level in Uzbekistan will drop a notch."

"I wouldn't count on it. From what I hear any shakeup in the status quo is liable to start a war. The War of the Aral Sea, sparked by a wide-eyed Canadian."

We laughed at that prospect, then gazed again at the sodden winter gardens.

"Hampstead Heath looks like Narnia still under the spell of the White Witch," I said. The fairy-tale landscape was making it hard to believe that I would be back in Tashkent in a week. "Maybe you've forgotten our way back through the wardrobe into the real world?"

"No such luck. Besides, I'm not sure my White Witch is preferable to your Mr. G. Now what do you fancy to eat? I know a Georgian place just up from Finsbury Park station, if it hasn't gone out of business. Or there's Jack Straw's Pub just over those snowy hills. My vote's for the pub."

"Robert, Merry Christmas and Happy New Year!" began an e-mail rom Ira that I received on December 29. "I hope you are doing fine and don't remember about our nice project. I will tell you briefly about the situation. I was called to the police the whole week. All our people including our part-time interpreters had been there already. The police are digging down and investigating not only the murder case but also our internal relationship with IFAS and our financial situation. I took away all the financial papers from the office – I destroyed all U.S. dollar receipts, including all the receipts for salaries. All the other financial papers are hidden in a secure place. If they ever ask about these receipts, you took them to Paris when leaving as it was the end of the year and all the financial papers had to be submitted to Paris. Robert, please, remember, nobody was getting salaries in U.S. dollar – NOBODY, no payment was done in U.S. dollar – everything in *sum*. I consulted a lawyer and she said that if the detectives find our receipts in U.S. dollar – signed by us – all of us will go to prison. Payments in hard currency are strictly prohibited in the territory of Uzbekistan, and trespassers bear criminal liability. So my official story is: I am not an employee of BDPA; I work for – [another company]. I was helping you and was making translations from time to time, and I was getting a bonus for that – in SUM. No dollars. All the employees were getting salaries in *sum*. The money was changed in the bank. If they ask for the bank receipts – they are in Paris. The police are waiting for your arrival – you willbe asked lots of questions regarding the financial relations with the project. I am sorry for this not very nice message."

"All our financial records!" shouted JC Torrion, shaking his head. "Why did she do this?"

We were sitting in his office on the eighth floor of a building on the rue Louis Vicat, a street that ran parallel to the Peripherique at the Porte de Brançion in Paris. It was Wednesday afternoon, January 3, 2001. That morning I'd taken the Eurostar from Waterloo Station under the channel to the Gare du Nord. Across from the station I'd had a croque monsieur and a draft Leffe and then taken the Metro to Porte de Vanves and walked the few hundred metres to JC's office.

"Maybe we should consult a lawyer," I suggested. "Phil Malone gave me the name of someone, a Canadian working for an American firm in Tashkent. Maybe we should call him right now."

He lit a Gauloise and thought for a minute. We were sipping espressos.

"It's not illegal to pay in dollars. This is how it's done all the time. It's right in our contract. This lawyer that advised Ira is wrong." He made a sarcastic snicker. "These lawyers are as corrupt as the government officials. If we call this lawyer, he'll just tell us to pay everyone off, especially him."

I looked out the window at a sea of high-rises that looked discouragingly like they could have been in Tashkent. Except here there were no cotton motifs. The ring road roared below, circling Paris like a medieval moat, separating the society of croissants, baguettes and cafés au lait from the encroaching suburban Third World transplanted from the Maghreb, Algeria and Ivory Coast.

"Ira's scared," I said. "Our whole team is scared. The police are just using a little blackmail to frighten them into talking. Shakhlo used the same tactic on Akhror to intimidate him. And I'm sure Bozov and Mr. G are saying anything they can to implicate us. They're carrying their game right over into Shakhlo's murder."

"Shakhlo's murder is not our business," said JC. His nose twitched in disapproval. "Our only business is getting paid!"

I knew that he was hanging onto David Pearce's promise in late October that Mr. G would pay up in the end. But Pearce had not responded to my last few attempts to discuss the situation, and with Anatoly now clearly working for Mr. G, I had little hope that the World Bank would, or even could, do anything.

"Mr. G's got the World Bank by the balls. Anatoly and Pearce anyway. It's going to come down to what the review mission says, and then what they actually do."

I finished my espresso and smiled at JC.

"So what if? What if we *had* paid off Mr. G at the start? Do you think he would have ever let us do our work?"

JC snorted in disgust. "This project was a joke from the beginning! Never was there any cooperation. Never was there any chance for us. Only games. I should not have signed this contract. If we have to go to court to get paid, we say that *never* was there good faith on their side! Never!"

It was the most upset I'd ever seen JC. He dragged on his cigarette and butted it out aggressively. Then he smiled a gloomy smile.

"But of course if we go to court we lose. It's Uzbekistan." He laughed bitterly. "I can't predict anything with this project any more. We just have to go back there and see what happens."

A banging at the front gate jarred me awake; the upper latch was open and it was rattling. I lay in bed half-awake thinking: I forgot to secure the upper bolt when I arrived from the airport last night. Then the doorbell began cooing relentlessly – the wild pigeon sounded frantic, insane. Then more banging on the metal gate.

I sat up and looked at the clock: 9:04 a.m. It was just after 5:00 a.m. in Paris, no wonder I felt groggy. The day before I'd flown from Paris to Frankfurt, changed planes and arrived in Tashkent at 3:30 in the morning. It was January 6. I drew back the satin curtains and peered out at the front gate. A naked cherry tree blocked part of the view, but there was no car parked in the driveway and I couldn't see anyone. More bangs and coos. Then a plaintive voice, "Robert! Robert! It's only me! Let me in!"

It was Ira. As I pulled on some jeans I glanced out the window again and saw a man step back from the gate into view. He was wearing a dark coat and a round-brimmed hat that made me think of Sherlock Holmes. I couldn't see his face but I knew who it was. I froze and my heart raced for a few seconds. Then I went to the kitchen and made coffee. I didn't answer the door and after a few minutes the banging and cooing stopped.

The coffee started to wake me up. When I finished it the phone started ringing. After about a dozen rings it stopped. But a few minutes later it started again. This time I picked it up.

"Why didn't you answer the door?" Ira asked, annoyed.

"Because you were with Detective Karimov."

"Robert, that wasn't Karimov! That was your neighbour across the street – Akhror's brother. He said he heard a car in the middle of the night and thought you were back. Robert, are you all right?"

"I'm okay. Just a little tired. Listen, everyone is okay. Your dollar salaries are not illegal. There's a clause in our contract that specifies that we pay all our staff in U.S. dollars. I'll show you if you like."

Her silence suggested that she didn't believe me and that she was more annoyed than relieved; I'd just insinuated that she'd erred in her destruction of our receipts and Ira did not like to be in the wrong.

"How are Sabit and Victor?"

"I don't know," she answered coolly. "I haven't heard anything about them." After another silence she said, "Detective Karimov wants to see you again as soon as possible." Another pause. "Robert, I can't come to the house any more. The police are watching me, and I'm sure

197

they're watching you. They're calling me every day, even on my mobile. I don't know how they got the number."

The wild-pigeon doorbell was cooing again, but less frantically this time. It was now early afternoon. I peeked through a crack in the gate.

"Sabit! You're not arrested!"

"Robert! I have so many things to tell you!"

I opened the gate and we hugged. His eyes were blazing.

"Tea!" he shouted. "First we need tea!" He charged past me into the kitchen, lit the gas stove and put the kettle on the flame.

"Robert, why didn't you buy tea? Look! There is hardly any left. This is a catastrophe!" A big ironic grin. "Follow me!" he ordered and I followed him out onto the patio.

I sat on a plastic chair and looked at him. He stood under the trellis. The grapevine was trimmed back, naked and brittle; it framed him like a stage prop. He pulled his fringe of beard. He looked like he was about to perform a soliloquy. He *was* going to perform a soliloquy.

"They are such terrible liars!" he shouted. But he was grinning. "Lies, lies, lies! All of them are liars! Drunken Tulenbai! That idiot Abbazbek! Those terrible comedians Bozie and Nadir! That devil Guiniyatullin! That little World Bank storyteller Krutov! Even our own Ira and her friend Boris the spy! Lies, lies, lies! And then more lies! They all lied to the police, they all told stories that said that *we* were guilty, that *we* were stealing from the project, that *we* were guilty of murder! Lies, lies, lies, lies, lies!"

He started pacing in front of me like a caged lion, a grinning caged lion. The courtyard was cool and damp, but pleasant enough. Tashkent's winter had still not arrived.

"Sabit, what happened with the police?"

He stopped pacing. He shook his head and clicked his tongue.

"They arrested me at one o'clock in the morning on the night you left. At first I refused to leave my flat because my wife was crying. Oh Robert, you don't know how terrible it is until your wife is crying!" But there was a sparkle in his eye, as if he was secretly pleased with her performance; she loved him. "I waited in Karimov's office in the Chilanzar police station all night. It was so humiliating! Finally I was taken to another room and Karimov and another detective interrogated me: 'Why do you wear a beard? Are you a Wahhabi? Were you involved with Shakhlo? Were you stealing money with her? And with Ferguson, the foreign specialist? How much money did you steal?'"

He threw back his head and roared, exposing the gap in his bucked front teeth.

"Can you believe Karimov wanted to know if I was having sex with her?" he was yelling now. "With that wicked prostitute!" He made a mock scowl, aping one of Nadir's faces.

I was spellbound by his performance. And wondering what the neighbours were thinking. Luckily they couldn't understand English.

He pulled up a plastic chair and sat down beside me.

"They hit me a few times," he confided. "But when I told Karimov that I was a poet, he told them to stop." He smiled. "They kept me at the station for four days! In the end they were only asking me one question: 'Where is Victor Tsoy?' Karimov still believes Victor is the murderer. I'm positive that Guiniyatullin and Bozov told Karimov that Victor did it."

"Where *is* Victor?" I asked Sabit. He stared blankly back at me. As in December, Victor's mobile was not available and he hadn't returned the message I'd left on his machine that morning.

"Robert! He must be at the Chilanzar police station!"

Sabit jumped up and ran back into the kitchen. Seconds later he was back with a teapot and two bowls. He set them on the table and sat down, letting the tea steep.

"But Detective Karimov is not so bad," he said, grinning at me. "He told me about his interview with Bozov. He said, 'This man is really *nothing.'* He said Bozov was trying to show off, saying that BDPA foreign specialists were *terrible*, that they had no new ideas!" Sabit roared. "Not like our Bozie! He has such *wonderful* ideas!" He poured tea into our bowls. "Karimov and I laughed. Yes, we even laughed! He told me all about Bozie's performance. He could see how simple our poor old Bozie really is." Another roar. "But Robert! I think he has no idea who killed Shakhlo."

I gave him a devious smile. "I have a few ideas."

"I know already that you are an excellent detective. You found the red van!"

"Are you ready?" I asked. "Here goes: Suspect 1: Faizillo, Shakhlo's 'nephew.' As we know Shakhlo and Faizillo were lovers. She paid him to drive her around and keep both her and the Lada serviced. He had motive: When Frank #1 came on the scene Faizillo must have been devastated – suddenly no more job, no more sex, no

199

more lover. And he had a fierce temper. Frank told me about it one evening in September, when things had settled down a little after Shakhlo's firing. Apparently Shakhlo had told Faizillo that he couldn't see well enough to drive her car any more – his eyes were black from a brawl. She fired him and he took it out on Frank. He shouted that Frank was a friend of Ferguson's and couldn't be trusted. Then Faizillo tried to hit Frank, but Shakhlo restrained him. 'That young guy is really crazy!' Frank told me. It was a good story: Faizillo assuming that Frank and I were friends and more of Frank's naivety – I'm sure he hadn't a clue Shakhlo and Faizillo had been lovers. Of course, I didn't say anything."

"Yes!" shouted Sabit. "She was leaving him. He was losing everything. And he was a hothead."

"But, Sabit – some big buts. Why has he been cooperating with the police? Is he really smart enough to cover up his crime by leading the police off on a wild goose chase? And he does seem to have a solid alibi. Although he could have taken the kids somewhere, dropped them for an hour, done the deed, and got them again and gone back to the flat and looked shocked at the bloody mess. But who was the accomplice? A girlfriend? A sister? I have my doubts Faizillo pulled it off. He would have to be a pretty clever guy to have managed all that. And besides, crimes of passion are not usually that carefully planned."

Sabit shook his head and clicked his tongue. Like my theories on why Shakhlo had cleaned out the safe, he was lapping it up.

"Suspect 2: The ex-husband. He was sitting next to Faizillo in the police station that day. He had a motive: Shakhlo was leaving, taking away his kids. And he may be destitute. Frank #2 met him and said he seemed to be a lost soul – no job, no drive. Which means he probably didn't do it. Lost souls don't usually express passion. Haven't got it in them. Too much self-pity."

"No, of course it's not him," agreed Sabit.

"Suspect 3: Victor Tsoy."

"Robert! You can't be serious!"

"Sabit, we must consider every possibility. As we know, Victor was not in Tashkent. He drove to Bukhara the day before the murder to attend an NGO conference. Let's say that he met up with some of the attendees that evening and set up his weekend alibi. Then around midnight he drives back to Tashkent. Saturday morning he calls Shakhlo and says he has a private matter to discuss with her. She sends the kids off with Faizillo and Victor and his accomplice arrive at Shakhlo's flat. Who is the accomplice? I don't know. A girlfriend –

I noticed that women tend to like Victor a lot. They butcher Shakhlo and leave. He drives back to Bukhara, hooks up with his friend again and makes it seem like he's been busy in Bukhara all weekend. Motive? That's easy. He blamed her for losing his job on the project. But much worse, she planted that letter in his employment file, damaging his reputation. Victor is very proud. And he really despised her."

Sabit's eyes grew large. "You don't really think so!"

"No, I don't. Victor is too moral. He believes inflexibly in always doing the right thing. That's why Mr. G wanted him off the project. Although," I added slyly, "I've noticed that highly principled people are sometimes very good at rationalizing their immoral acts."

Sabit shook his head and clicked his tongue. "Impossible, Robert! Not Victor!"

"Suspect 4: Kadirbek Bozov."

"Not poor old Bozie!" Sabit grinned and looked very pleased.

"JC told me that during the contract negotiations he thought Bozov and Shakhlo were lovers. I know it seems far-fetched, but Shakhlo did seem to have wide-ranging sexual tastes. Other than Faizillo, who was really just a toy boy, her lovers all provided her with opportunities. Bozov got her the job with us. Possibly he was taking a percentage of the money she was skimming. Or maybe he wasn't and *that* was the problem. He felt he *deserved* some of our money and when she didn't hand it over he had her sliced up. It's also possible that he really loved her and was jealous of Frank #1. But Bozie in love? Sounds preposterous. Though we do know that our poor old Bozie is not a brave soul; he would never have been able to do the slaughter himself. But he likes to think he's a bit of an operator and a contract killing is within the sphere of his limited capabilities."

Sabit hooted again. "No, he couldn't do it! It's not possible for poor old Bozie!"

"I tend to agree. Okay, suspects 5 and 6: Sabit Madaliev and Robert Ferguson."

"No, please!" shouted Sabit.

"Saturday morning, if you recall, we were here together, going through that huge pile of receipts, listing them all in a financial report for Bozie so that Mr. G might pay BDPA. It took us till the early afternoon. Or did we? Maybe we got talking, as we do, thought about Shakhlo and her lies, her tricks and her pilfering, got all worked up and decided that she was the linchpin in the works, the source of our humiliation. Bozov is too stupid,

Mr. G too remote – Anatoly once claimed that he didn't know what's really going on with the project. And so it must be Shakhlo. She was behind all our problems. So we jumped in a car and went over there and slash, slash, slash." I frowned. "I'm sorry to have to say, Sabit, that your slashes were rather weak, a little bit *lady-like*."

"Robert, you're completely crazy!" howled Sabit. "You should never have come to Central Asia. It has completely destroyed your mind!"

"I'm not finished yet. Suspect #7: Mr. Rim Guiniyatullin. Mr. G runs a tight ship. He doesn't like it when a member of his crew is caught with his, or in this case *her*, hand in the cookie jar. Certainly not if a foreigner does the catching. Shakhlo had to be punished. She had overstepped the line, stolen blatantly rather than shrewdly as she should have." (Soheil Ramanian, the go-between, had told me that in Uzbekistan, unlike Azerbaijan, corruption was kept behind the scenes; it was never openly discussed.) "Worse, she was unforthcoming when he asked for his usual cut. Then she tried to escape, flee to Brussels with Frank #1 and the loot. She didn't quite make it. Mr. G had her taken out. But of course the old devil is clever. He hired a woman to kill her to make it look like a crime of passion, just to put the police on the wrong track."

Sabit's eyes gleamed. "Yes, it is possible!"

"All right Sabit, the last one. Suspect #8: The property agent. Last April, when we rented this house, a *mafiya*-type character showed it to us, and I remember he treated Shakhlo with contempt. Later Akhror told us that Shakhlo – or indirectly BDPA – had paid this man a fee, his commission or kickback for finding the house. Then, just couple of weeks ago, Akhror told us that this same agent was demanding another fee because soon we would be leaving. But I refused to pay and told Akhror that he should ask Shakhlo for it. I said something flippant like, "She took enough money from us, so it's up to her to pay this sleazy guy off."

Sabit's eyes grew wider. "You don't think ... "

"But if Akhror told the agent that she was skimming money off the rent each month – and as we know, Akhror is the type to say something like that – then maybe the agent decided that the money she'd stolen was rightfully his. He was the one who had been ripped off. And last Saturday morning he went to Shakhlo's flat to get back his money. And when she wouldn't pay up, he had his assistant, chop, chop, chop."

Sabit was thunderstruck. Slowly his eyes widened in comprehension.

"You're right!" he shouted. "*He* is their prime suspect!"

Bozov was glowing. So was Nadir.

"*s'Novym Godam*, Robert!" they shouted. "Happy New Year, happy New Year, happy New Year!" They both shook my hand vigorously. Bozov complimented Ira in Russian and Nadir interpreted: "Such a beautiful woman, like a flower in the Kyrgyz mountains!" Ira blushed and then looked insulted.

"Please to sit down!" shouted Nadir.

"Tea!" shouted Bozov.

"I am preparing tea!" shouted back Nadir.

When we were seated, Bozov lit a cigarette and leaned back in his squeaky chair.

"Where is my good friend Torrion?" he asked playfully. I told him he would arrive in a few days. "Where is our equipment?" he demanded. "You must move everything into my office!" He pointed at the empty corner that had been waiting for "Company BDPA."

"*Everything* you must move here!" shouted Nadir from across the room where he was making the tea.

According to our contract, all our office equipment and supplies were to go to Component B when our work was completed. It was the standard arrangement on most international projects: when the foreigners left, the locals got their goodies. They wanted our computers, which were newer than the ones they were using, and Bozov had a particular interest in the digital videocam. But I had no intention of handing over a thing until Mr. G paid BDPA.

"*Kanyeshna*," I said – of course. *Kanyeshna* had become my standard response to most of his demands. "Why don't you give me the forms and Ira and I can begin the process?"

"Robert," said Bozov with stern authority. "I've appointed a property transition committee. You will meet with them tomorrow to make all agreements and plans."

It sounded very Soviet, and very unnecessary. I could see how determined he was to get our things. I glanced at Nadir, expecting his eyes to roll, but he frowned at me and poured the tea.

"A committee is a good idea," I said, smiling at Bozov. Then I leaned forward and said confidentially, "When Mr. Torrion gets here in a few days, he would like a private meeting with Mr. Guiniyatullin. Just the two of them and a translator."

Bozov's eyes lit up. "A private meeting?" He eyed me for a few seconds, his usual smirk dissolving into an excited smile. *This foreign bastard is stupid, but maybe he is finally learning!* "What is the topic?"

203

"The topic is a 'gentleman's agreement.'"

"Of course! We will arrange everything!" Bozov was overjoyed.

"Whoa! Whoa!" Nadir was grinning. "You see," he said smugly to me, "finally you and Torrion understand how to work in Central Asia." Then he looked at Ira. "Whoa, whoa!" he shouted again. But she gave him a stony look. And he responded with a mock scowl, like she was such a poor sport.

"Robert, I'm so sorry I couldn't reach you. My mobile is dead – I must pay the bill!" Victor was in a relaxed mood.

It was the following morning. Ira, Mila, Sabit and I were packing up our equipment at 45 Tazetdinova Street. We were going to store everything until things were settled. Victor's arrival was a surprise as I hadn't been able to reach him.

"Tell us what happened with the police?" I asked him. We were sitting in the living room surrounded by open cardboard boxes and stacks of file folders and reports.

"I went to the Chilanzar station the day after I met you, Rob, at the Salty Dog. Detective Karimov questioned me for several hours. It was easy to defend myself. You see, he had nothing on me. He had nothing that tied me to Shakhlo's murder. He had only some stories that tried to blame me. Soon he was telling me his theories on the murder."

He paused and smiled self-assuredly as he puckered his moustache.

"You see," said Victor, now in his analytical persona, "Karimov is a clever man. But he has no evidence. He has no murder weapon. He has no fingerprints. Or witnesses. He only has a lot of statements and he told me that many of them contradict each other. He is having problems solving this murder. He knows his job, but this case is a mess."

Sabit gave me an excited glance.

"He mentioned Bozov," continued Victor. "He said that this man had no integrity. He called him a clown. He talked also about Faizillo, Shakhlo's mother, Tulenbai and Abbazbek. He didn't mention Mr. G. But he said that everyone seemed very determined not to tell the truth. He said he didn't know what had happened on the project, but that he felt there was some kind of conspiracy going on. I felt sorry for Karimov. I don't think he'll ever find Shakhlo's murderer."

On Friday, January 12, JC had his private meeting with Mr. G. I waited for him in Bozov's office with Ira, Bozov and Nadir. We were all in happy moods, each of us for different reasons.

Bozov was happy because he believed that JC was in Mr. G's office finally agreeing to make a gentlemanly payment and wrap up our contract. We had failed and we were leaving. He had won and we were going to pay a big penalty for losing, as we should. He had proven that they didn't need any foreign specialists. He would be a hero among his cohorts: Valentina, Bayalimov, Talbak, Tulenbai and Abbazbek. And probably his peers, the heads of the other four components, who would slap him on the back. Mr. G would stop growling at him for a week or so. And now they could get back to doing what they were doing before Ferguson had arrived and screwed everything up.

Nadir was happy because Bozov was happy.

Ira was happy because she was relieved. She was finishing her contract with us that day; her ordeal was finally over. She'd told me that morning that soon she was leaving for Europe for a holiday. It would be her first trip to the West.

I was happy for three reasons. First, I was thinking: It's almost ended. I'll never have to endure this again. As I sat drinking Nadir's tea – without lemon – listening to Bozov tell me another preposterous anecdote, I was feeling all those tricks and scams and rebukes floating up over my head and dissolving into the smoky air. A huge load was lifting off my shoulders. Second, I was happy because we'd notched up a last triumph: we'd escaped Shakhlo's murder investigation unscathed, despite the efforts of Bozov and Mr. G. And third, even though we'd failed in our mission, even though JC was at this very moment caving in, agreeing to pay Mr. G so that Mr. G would pay him, I was happy because we'd survived. We'd battled the forces of repression, corruption and evil and endured. *Ours was the moral victory!* (But in Central Asia, moral victories have a way of counting for little in the overall scheme of things.)

"The World Bank has declared the Aral Sea project unsatisfactory," said JC, smiling as Sabirjan zoomed us down Navoi Street towards Mafia Pizza. "He showed me an official letter in English. It said the project was unsatisfactory because it was failing to meet the goals of improved water management in the Aral Sea basin. His project is failing! Mr. G is very upset! He is very sad! He wanted my *sympathy*!"

"Poor Mr. G!" I said.

Sabirjan pulled a U-turn and came to an abrupt stop next to a new Mercedes with tinted windows that was parked brazenly in front of the restaurant. The owner's I thought; *biznes* is good.

"But what about your 'gentleman's agreement?'"

JC laughed. "Maybe now we don't need this!"

Mafia Pizza was busy but we managed to find a table by a window that overlooked the street. A large TV mounted high on the wall was blasting an Uzbek music video – four skinny young men in dark suits and ties were singing about their broken hearts in a cherry orchard in full blossom. The mournful wailing sounded like a cross between Turkish and Indian music. The owner, sitting at his usual table discussing biznes with his cronies, recognized me and called over, *"Asalam aleykum!* Hello mister!" I grinned back and nodded, realizing that there were a few things in Tashkent that I was going to miss. He sent over a pot of tea on the house. It came with lemon.

"I asked about the invoice," said JC after we ordered Chimkentski beer and pizza from the waitress. "But Mr. G was so dejected! He didn't want to talk about it!" JC couldn't stop smiling. The beers arrived and JC held up his glass.

"To Mr. G," he said. "To his failure at water management." We tapped and drank.

"But this was strange," JC continued. "The letter didn't mention Component B at all. It mentioned some of the other components as being ineffectual, but ignored public awareness. And Mr. G told me that this component was working well."

"That's because Anatoly Krutov never gave them my report. He and Mr. G made sure that the review mission ignored our situation."

"I wonder what the World Bank will do now," said JC.

"Let's go see Anatoly and hear his version of things. For old time's sake."

"It's so very complicated!" said Anatoly with his chirpy smile. "The review committee found that the project was not effective in several areas, especially in its role in water management in this terrible drought." He shook his head and looked grim. "It's very bad. And the political situation is only getting worse. The bickering is terrible! If the project doesn't keep everyone talking, who knows what will happen?" He shook his head. "I don't know what Washington will decide."

"What will happen with Component B?" I asked.

Anatoly revived his sunny expression. "This component is not so bad. Really! Robert, believe me!" He winked.

"Anatoly, why was José Bassat told not to come? Why was I locked out of the review process? Why did you keep my confidential report from the review but give it to Mr. Guiniyatullin?"

"You make it sound like we were *throwing you to the wolves*!" said Anatoly with a big smile. But he dropped it when he saw we weren't amused. "Jean-Charles, Robert. Mr. Guiniyatullin is only one difficult man. We have five governments that are full of Guiniyatullins. Central Asia is full of Guiniyatullins. They're everywhere! So many things go on here that are impossible to explain. It's all so very complicated."

He shook his head and looked dejected, as if worn down by the weight of all those Guiniyatullins all glowering and growling at him. Then I saw them, thousands and thousands of them, walking the streets of Tashkent in a huge crowd and cheering in unison as President Karimov cut a ribbon to open yet another new development, and in massive standing ovation at the Alisher Navoi Theatre for the Uzbek opera "Timur the Great." Mr. G cloned. Mr. G everywhere.

"Robert, your report had a deep effect on Mr. Guiniyatullin. I knew it would. That's why I gave it to him. It made him realize what was really going on with Component B, with Bozov and the team leaders." He sighed. "I really don't know what will happen with this component now."

"Anatoly, are you telling me that Mr. G had no idea what Bozov and his cronies were up to? How many times did you tell me that Bozov was nothing, that he was only Mr. Guiniyatullin's soldier?"

"Robert! Mr. Guiniyatullin is very busy. He has four other components. He has the IFAS committee to answer to. He has four national governments who are all against Uzbekistan. Believe me, he can't keep *his finger in every pie*!"

"What about BDPA's invoice?" asked JC.

"Jean-Charles, your contract is finished!" Anatoly was sunny again. "Now it's time to pay up."

But he didn't suggest who should pay first.

20   A Gentleman's Agreement

I never made it to the Aral Sea. I never gazed across its lingering waters from one of its emaciated beaches. I never waded barefoot along its brackish shore or soaked up any of its dwindling but possibly still potent energy. I wonder whether doing so might have kept my vision clearer, helped sustain me and allowed me to fend off the tricks, the scams and the demons. (Or whether it might have cast a spell on me, corrupted me, ordered me to take shelter under the monster wing of Mr. G.) I'll never know. Bayalimov claimed there was nothing to see anyway, and probably he was right; he turned out to be surprisingly truthful about many things.

So my image of the Aral remains the desolate patch of shimmering blue-green in the midst of Kyzylkum Desert that I saw out the airplane window as I arrived. Because I never got there, my Aral remains mostly in my imagination, which is maybe more satisfying – like reading the book but not seeing the movie (as we know, the book is *always* better than the movie). But my idea of it, and the water crisis, is much richer now. It's been coloured by Talbak's Mona Lisa of the Aral, Nazarov's revived decrees to stop the spitting in canals and Bayalimov's warnings from the Qur'an: "See you the water which you drink? ... Were it Our Will, We could make it salt (and unpalatable)." My favourite is the legendary Alisher Navoi poem "Farkhad and Shirin," a scene from which is illustrated in ceramic tile on the wall of a Tashkent metro station. One day Sabit told me the story: a beautiful princess is auditioning suitors and in the end chooses the man who promises to build her the best canal. Uzbek practicalities. These are the recollections that hold that maddening and often bewildering experience together for me now.

But the picture I see first when I look back on my year in Central Asia is not the disappearing salty pools of the Aral Sea, the barren expanses of steppe and desert, or the tank parked in front of the Averso Hotel in Dushanbe, or the brutalist KGB building in Tashkent. It's not the glower of Mr. G, the gloats of Bozie and Nadir, the team leaders lined up across the table chastising me or the young soldier crying into his empty mug in the Czech Beer Bar, though with time I realize that these are all memories I will cherish. No, what I see is the view from our table beside the pool at the Labi-Hauz in Bukhara. The afternoon sun had just sunk behind the yellowbrick walls of the Nadir Divanbegi Medressa and the light had turned everything golden. And then the pool erupts. It gushes and spurts and sprays Gabi and me and everyone else. Water. Ordinary incredible water.

Two days after I left Tashkent, JC Torrion met with Rim Guiniyatullin in his new office with the huge map of the blue Aral Sea covering one wall. Mr. G was back in wily form; apparently the negative review of the World Bank Mission only had a short-term effect on him. Not surprisingly they failed to reach a "gentleman's agreement." By then, JC was not inclined to behave much like a gentleman, and Mr. G ... well, he never had behaved like one. It would be 15 months before the invoice was settled and the deadlock that had suspended all public awareness activities under Component B resolved.

"I am concerned as you are with the fact that the problems with Component B seem to be minimized by both IFAS and the World Bank team," wrote José Bassat on January 27, 2001. "The fact that the teams have produced so little and of such poor quality is symptomatic of the problems existing with this component ... I am planning to travel to Central Asia in the spring and will have to be quite clear about the uselessness of dedicating more resources to this component without a real change in attitude and approach on the part of our counterparts. In any event, I want you to know that I am very grateful for your commitment and for having held on through the difficult times in Tashkent. I am very much aware of the obstacles which were in your path and which prevented you from delivering what you had planned to deliver."

In late January 2001 Frank Thevissen, Frank #1, returned to Tashkent. Troubled over Shakhlo's murder investigation, he planned to undertake his own inquiry. Bozov, reportedly in very friendly spirits, shanghaied Frank into providing some input on the national teams' public awareness strategies. (Apparently Frank was never paid for his efforts.) During this trip he met the daughter of Ludmila, Bozov's submissive red-haired secretary, married her and took her back to Brussels.

Ira got her wish to escape Uzbekistan. "I've been in Moscow for half a year already," she wrote in July 2001. "I'm working for Rick Flint [one of our foreign specialists] on an information and communication project funded by the European Union. I'm doing press releases for the project's events and arranging media coverage ... Frank Thevissen married Ludmila's daughter! I tried to see him in Brussels when I was there in June. He had asked for the meeting, but then he stood me up. Just like Frank."

Phil Malone helped Victor Tsoy get a job as a local project manager on the World Bank-funded Tian-Shan Project. Victor also continued his work with his NGO, Rabat Malik. Phil, Frank #2 and Rick Flint all continued to work with their communication firms on projects in the former Soviet states.

Boris Babaev launched a newspaper. He worried that the publication of this book would alert Uzbek state TV to the fact that he was actually receiving two salaries when he worked for BDPA. I'm so sorry Boris.

On September 11, 2001, Sabit Madaliev boarded an Uzbekistan Airways flight in Tashkent bound for New York. Because of the terrorist attacks on the World Trade Center, the plane was diverted and he ended up in Goose Bay, Labrador. He called me from there: "Robert, I am in your country!" Serendipitously, I was about to leave for Newfoundland, to Gros Morne National Park, about a thousand kilometres south of where he was stranded. But he was back in Tashkent by the time I got there. Two weeks later he e-mailed me from a small town in upstate New York where he was attending a Sufi retreat. But before I could drive down there to visit him he was back in Tashkent, reporting that it was not a good time to be a Muslim in the United States.

In the spring of 2002 the Uzbeks held an Aral Sea Forum in Tashkent to face up to their escalating water woes. Drought, excessive water use, population growth and cotton production were all continuing unabated. Their solution? To resurrect the old Soviet plan to divert the Ob and Irtysh rivers in Siberia south to the Aral Sea. "There is no other way to address this problem but to source water outside the region," said Ismail Jurabekov, an aide to President Islam Karimov. "Siberian water would help us grow fruit, vegetables, cotton and grain crops, the bulk of which would feed Russian provinces." The Russian ambassador to Uzbekistan, Dmitri Ryurikov, also attending the conference, said diplomatically: "Moscow is unlikely to agree in a hurry."

However, in December of that year the plan got another shot in the arm from an influential Russian politician. Moscow Mayor Yuri Luzhkov sent an official letter to President Vladimir Putin reasoning that the Siberian water diversion scheme would allow Russia to sell its excess fresh water to Central Asia, and prove very profitable. Luzhkov suggested funding the proposed US$12–20 billion project with loans guaranteed by future sales of fresh water to the five former Soviet states. But some Russian experts argued that the waters of the Ob are too contaminated by oilfields to be used for irrigation. And Nikolai Glazovsky, head of Moscow's Institute of Geography, told reporters: "Such an idea should not be nurtured by any normal-thinking person."

Meanwhile, closer to the disaster, the Uzbek government has been turning a blind eye to the thousands of Karakalpaks who have been bribing officials to get across the border into Kazakhstan. According to a correspondent in Karakalpakstan, between 1995 and 2002 an estimated 250,000 people, a sixth of the "autonomous republic's" population, fled their homeland as crops failed and health problems continued to mount. The Uzbek government is doing little to alleviate Karakalpakstan's plight, but has a strong strategic interest in the territory, which makes up one-third of the nation's area. And it seems to have found a way to manage the disaster zone: empty it of people.

In October 2002 the United Nations Environment Program announced that the Aral Sea would no longer exist by the year 2020. "There is little that can be done at this stage to save the sea from extinction," said a UN expert, adding that the disappearing sea was a symbol of the failure of the five Central Asian states to cooperate on vital regional issues. This announcement followed a meeting of Central Asian heads of state in Dushanbe, Tajikistan, under the rubrics of the Central Asian Cooperation Organization and IFAS, the International Fund for Saving the Aral Sea. Observers commented that there was little cooperative spirit: "Predictably, Central Asian leaders asserted their own geopolitical interests rather than working towards a joint solution."

News from President Islam Karimov's website in December 2002 provided results from "the authoritative independent U.S. Pew Research Center's 2002 Global Social Relations" survey of 44 countries. The results showed that President Karimov had "the highest rating by a country's citizens of its president." Karimov scored 95 percent; Russia's Vladimir Putin made 85 percent; George W. Bush of the United States was at 71 percent and Britain's Tony Blair, actually a prime minister, made do with 54 percent. Further, 62 percent of Uzbeks surveyed said their "current economic situation" was very favourable and 69 percent were "satisfied with life." By comparison, just 56 percent of Canadians, 41 percent of Americans, 32 percent of Brits and

20 percent of Russians were happy with their lives. The site clarified the results for us: "This survey ... shows the correctness and success of Uzbekistan's independent path of development and the support of the people for the reforms of President Islam Karimov." There was no indication of the survey's margin of error.

President Niyazov of Turkmenistan – "Turkmenbashi" – has been busy with the *really* big issues of time and life. He has renamed the months of the year after national heroes. January is now called "Turkmenbashi" and April is named for his mother. He has also divided life into 12-year cycles. Childhood officially lasts now until age 12. Adolescence continues to age 25. Turkmens aged between 25 and 37 are youthful, while those aged between 37 and 49 years are mature. The next 12-year cycles are classified as prophetic, inspirational and wise. Old age only begins at age 85, which is a relief – I think he might be onto something. According to the World Health Organization, the average life expectancy at birth for Turkmen men is 60, and 65 for women. The new decree saves the vast majority of the population not just the humiliation of old age, but the burden of being inspirational and wise. Niyazov turned 62 in 2002. He has entered his inspirational period, which is a surprise to no one.

JC finally sent some substantive news on his outstanding invoice on January 21, 2002, almost exactly two years to the day I first arrived in Tashkent: "Apparently Mr. G has been in a very uncomfortable position since we left. The World Bank suspended the component and all five national team leaders were fired. They very strongly criticized Mr. G. IFAS headquarters in Turkmenistan has also been very hard on Mr. G. The World Bank in Tashkent is proposing that IFAS resume Component B – there is still $1.6 million (US) in the component budget – which should also include payment to BDPA ... It seems that our payment is the World Bank's condition to resume the component."

On February 6 he reported: "It seems that we might, finally, get out of this ordeal. Mr. G is a bit trapped. According to a report from our middleman [JC had hired a local lobbyist to negotiate the final payment], Mr. G became the bad guy in the region because the component has been suspended and has no money. Mr. G hates BDPA. Mr. G hates JC Torrion. Mr. G hates Rob Ferguson. Mr. G hates the World Bank ... It seems that the World Bank would be OK to resume financing Component B if BDPA is paid ... Anatoly Krutov has suggested something as a way to 'save' the pride of Mr. G."

Then on April 19: "At last! Negotiations were quite hard. My lobbyist is convinced that Anatoly Krutov is 100 percent pro Mr. G. The result is far from what we could expect. BDPA will be paid except for the period following my partial suspension and several items from invoices over the rest of the year as they keep saying that no report has been approved. Plus, of course, a significant "gift" – a very significant one! – for Mr. G's pride!"

Shakhlo's killers remain at large. Detective Numan Karimov seems to have given up on her case. "Davron Abdullayev," someone representing Shakhlo's family, saw a website promoting this book and sent an e-mail to my agent on June 18, 2002: "Yesterday we went to Uzbek foreign affairs secret intelligence services and to INTERPOL in order to bring an action against Mr. Ferguson and this website about this criminal act, and they will get in touch with the Canadian police authorities." It wasn't clear if the criminal act was Shakhlo's murder or the writing of this book. Or both. I still await the RCMP and INTERPOL, with trepidation.

Bruno de Cordier, the Belgian who entertained Gabi Buettner and me up in the mountains of Kyrgyzstan, was in Tajikistan for a short-term mission in October 2002. He wrote: "I am working with a Tajik journalists' NGO and the chairman actually worked for the Tajik branch of the infamous International Fund for Saving the Aral Sea. He told us that IFAS was as big a catastrophe as the Aral Sea itself. He knew the whole pantheon including Guiniyatullin, Bozov, etc., and he also knew about your book and the effect its summary had among some in Tashkent. One quote from our conversation that might interest you: 'This book is to be published in 2003. But I know that it will never be published. Guiniyatullin has very good connections with the World Bank. They will buy out the publisher, or pay the author so that he withdraws his manuscript.'" My publisher and I waited for the World Bank's offer with considerable excitement. But in the end we were both disappointed.

Rim Guiniyatullin remains at the helm of both the International Fund for Saving the Aral Sea and the Water and Environmental Management Project for the Aral Sea Basin. BDPA was paid, but contrary to the deal brokered by the World Bank, activities under Component B were not resumed. Instead Mr. G sent Kadirbek Bozov home to Kyrgyzstan and dissolved the five national public awareness teams. The media of Central Asia no longer runs stories on the great tragedy of the Aral Sea and the exceptional talents of the local

213

specialists of the Aral Sea project. Sabit reported that Nadir, the Aral Sea project's only nuclear physicist, was also the sole employee remaining in the nearly empty Component B office. No doubt he was chattering happily on the phone and offering cheerful grins and tea – "and *limon!*" – to anyone popping in for a visit. The project was expected to wind up in late 2003 unless IFAS could find donors to keep it going by extending its funding. I offer no predictions.

In July 2002 Mr. G sent an email to Jean-Charles Torrion demanding that he stop me from writing this book. JC wrote back: "Mr. Ferguson is no longer under contract to BDPA. He is free to write whatever he wants. I just know that Mr. Ferguson is an honourable and trustful person."

Of course I am. *Kanyeshna.*

Epilogue

It's been over a decade since I left Central Asia and the disappearing Aral Sea. "One of the planet's worst man-made ecological catastrophes," as it's frequently called, has continued largely unabated and the world is still mostly ignoring it.

The Aral is now less than ten percent of its original size and has drained into two mini-seas, a larger one in Kazakhstan and a lesser one in Uzbekistan. Thanks to a concrete dam over the Syr Darya completed in 2005 with World Bank funding, the North Aral Sea has been creeping back. The water level has risen, its salinity has decreased and fish have actually returned, partly reviving a long dormant fishing industry. Meanwhile the South Aral Sea, left to its unnatural fate, has dried up into a thin brackish strip at the far west of the former sea, greatly expanding the Aralkum, the huge toxic desert. Uzbekistan continues to use up the Amu Darya in growing its thirsty cotton crop, leaving no hope for this part of the desiccated seabed. And while the United Nations Environment Program's prediction the Aral Sea will disappear by 2020 now seems improbable, the Aral's long-term outlook remains extremely bleak.

As Mr. Guiniyatullin told Jean-Charles Torrion back in January 2001, the outcome of the Aral Sea Water and Management Project was "unsatisfactory." The final report, dated June 30, 2003, made that official, stating: "The project did not achieve the stated objective of reducing withdrawals of water for irrigation by 15% over the project period." With regards to our public awareness effort, the report said: "Component implementation faced difficulties from the start. The component objectives and design were extremely ambitious and did not consider the region's political realities. In retrospect, the aim of changing water users' behavior through creating public awareness of the urgent need to conserve water ... was unrealistic. The major causes of inefficient water use were and are dilapidated infrastructure ... and poor government policies. These are quite separate issues from inadequate public awareness. The component's design, based on a modern public awareness campaign aimed at specific groups, was premature in a cultural context where such campaigns are broadly understood as government propaganda, and where governments still dominated agriculture and irrigation."

Robert Ferguson

Real democracy has yet to come to Central Asia. In Uzbekistan Islam Karimov, re-elected to a third term in 2007 despite the Uzbek Constitution's two-term limit, should ensure his iron rule will continue until at least 2015. Colourful President Saparmurat Niyazov of Turkmenistan, aka Turkmenbashi, died of a sudden heart attack in late 2006 and was succeeded by Gurbanguly Berdimuhamedow, who rescinded some of Turkmenbashi's more idiosyncratic policies, such as banning opera and the circus for being "insufficiently Turkmen," and, eager to sell his nation's natural gas, has made overtures to the West. Mountainous Kyrgyzstan, always the more democratic, open, and to me appealing, of the five "'stans," has seen a decade of political upheaval culminating in riots between ethnic Kyrgyz and Uzbeks in 2010, a enduring consequence of Stalin drawing freely over the map of Central Asia with his pencil. Tajikistan remains repressive, corrupt and poor, and Kazakhstan, despite an international bruising following Sacha Baron Cohen's mockumentary *Borat: Cultural Learnings of America for Make Benefit Glorious Nation of Kazakhstan*, continues to aspire to be the Texas, or Alberta, of Central Asia with President virtually-for-life Nursultan Nazarbayev presiding over a mostly buoyant economy exceptionally dependent on oil and gas exports.

In 2005 my Belgian friend Bruno de Cordier relayed some tragic news he'd spotted on a website covering news from Central Asia: "Victor Tsoy, leader of the Slow Food Tashkent Convivium, died in a hiking accident July 28 while visiting a village in the mountains." Victor had hiked and guided visitors in his beloved Chatkal Mountains for years. All of us who knew him were distressed and suspicious. Why was he up there allegedly hiking alone on a Friday evening? Unfortunately Victor's good work with his NGO and his high principles had made him enemies, as I knew only too well from the way Mr. G had forced him from our team. Unhappily, like Shakhlo's murder, there is no way of finding out what really happened. As Sabit used to say regularly, "Robert, knowing the truth about things in Central Asia is impossible!"

As for Sabit Madaliev, in 2006 he wrote a book called *The Silence of the Sufi*, which he called "a spiritual journey, or stranstvie, a Russian word for wandering that is laden with literary, religious, and historical nuance." In 2010 he wrote me about his life in Tashkent: "You know it is just impossible to live here ... but I am not too much worried — God won't let me down."

216

References

My sources include my own notes and journal and hundreds of e-mails, memos and reports pertaining to the project and the geopolitical situation in Central Asia. The internet was also an invaluable resource, with information from sites such as uzreport.com, *Pravda Vostoka* and times.kg *(The Times of Central Asia),* Canadian, U.S. and British newspapers and organizations such as *Médecins Sans Frontières.* Eric Wahlberg in Tashkent sent me many "unofficial" articles, including those by "Karina Insarova," the pseudonym of a journalist living in Karakalpakstan. Of the following selection of articles and books, some served as specific references for individual chapters, while others provided inspiration and ideas, which to me was just as important.

Alaolmolki, Nozar. *Life After the Soviet Union: The Newly Independent Republics of the Transcaucasus and Central Asia.* Albany, NY: State University of New York, 2001. A comprehensive look at the emerging political realities of Central Asia.

Arnold, Matthew. *Poetic Works.* London: Oxford University Press, 1969. The last part of Arnold's epic poem, "Sohrab and Rustum: An Episode," quoted at the start of the book, is now a sad eulogy to the Aral Sea.

Bailey, F. M. *Mission to Tashkent.* Oxford: Oxford University Press, 1992. A true story and terrific spy thriller that unfolds in the last years of the Great Game; remarkable to me for revealing that the trickery and intrigue then continue in the same form today.

Bissell, Tom. "Eternal Winter: Lessons of the Aral Sea Disaster." *Harper's Magazine* April 2002. A thorough overview of the impact of the Aral Sea disaster on Karakalpakstan.

Blagov, Sergei. "Russian Water on Troubled Soils." *Asian Times* 18 Dec. 2002.

Brzezinski, Matthew. "Whatever It Takes." *New York Times Magazine* 16 Dec. 2001.

Byron, Robert. *The Road to Oxiana.* London: Picador, 1981. In 1933–34 Byron, an English scholar and aesthete, recorded this wonderful diary of his adventures in Persia and Afghanistan on a quest for the origins of Islamic architecture that was inspired by a photo of a Seljuk tomb-town on the Turkmen steppe.

Central Asia in Perspective, World Bank World Development Report 2000–2001. New York: Oxford University Press, 2000.

De Villiers, Marq. *Water.* Toronto: Stoddart, 2000. A convincing and absorbing overview of why water is emerging as the resource issue in this century.

Ditmars, Hadani. "Stuck in a Danger Zone and Plagued by the Old Soviet Bureaucracy, Uzbekistan Struggles to Fight Back." *Report on Business Magazine* November 2001.

Diagnostic Study: Rational and Effective Use of Water Resources in Central Asia. Tashkent: United Nations Special Program for Economics of Central Asia, 2001. Shortly before I left Tashkent I visited SANIIRI, the Central Asian Scientific Research Institute for Irrigation, and received a copy of this report, which pointed out two trends: that privatization and water charges in Kazakhstan and Kyrgyzstan were starting to reduce water consumption in those states; and that from 1995 to 1999 water use had declined in four states but not Uzbekistan, where it increased by 16 percent.

Encyclopaedia Britannica. New York: Encyclopaedia Britannica, Inc., 1963. Used as a reference for statistics on water bodies, topography and state areas.

Fadiman, James, and Robert Frager, ed. *Essential Sufism.* San Francisco: HarperSanFrancisco, 1997. This unassuming guide provided the background in the discussion of Sufism in Chapter 14.

Glantz, Michael J., et al. *Creeping Environmental Problems and Sustainable Development in the Aral Sea Basin.* New York: Cambridge University Press, 1999. A scientific overview of the causes and effects of the Aral Sea disaster.

Greene, Graham. *The Quiet American.* London: Vintage, 2001. A novel about naive optimism clashing with the cynical realities set in French-run Vietnam of the 1950s, a theme I could relate to.

Hopkirk, Peter. *The Great Game: On Secret Service in High Asia.* Oxford: Oxford University Press, 2001. Hopkirk tells a great tale set in Central Asia of the Great Game.

Kaplan, Robert D. *The Coming Anarchy: Shattering the Dreams of the Post Cold War.* New York: Vintage, 2001. Part travel, part philosophy and history, Kaplan serves up a provocative book that challenges conventional thinking on international affairs.

– *Eastward to Tartary: Travels in the Balkans, the Middle East and the Caucasus.* New York: Random House, 2000. More great stuff from Kaplan, touching on Central Asia.

– *The Ends of the Earth: A Journey to the Frontiers of Anarchy.* New York: Vintage Departures, 1997. A terrific section on Central Asia; he captures the mood, contradictions and frustrations that permeate the region.

Kipling, Rudyard. *Kim.* London: Penguin, 1989. Set in British India (and today's Pakistan), this classic is a study of two cultures, East and West, meeting in the era of the Great Game.

Macleod, Calum, and Bradley Mayhew. *Uzbekistan: The Golden Road to Samarkand.* 3rd ed. London: Odyssey, 1998. A literary travel guide, especially good on Samarkand and Bukhara.

Madaliev, Sabit. *The Silence of the Sufi: And I Do Call to Witness the Self-Reproaching Spirit.* Iowa City, Iowa: Autumn Hill Books, 2006.

Mayhew, Bradley, Richard Plunkett and Simon Richmond. *Lonely Planet Central Asia.* 2nd ed. Melbourne, Australia: Lonely Planet Publications, 2000. The most complete guide on the region, thorough, accurate and refreshingly direct.

Meyer, Karl E., and Sahreen Blair Brysac. *The Tournament of Shadows: The Great Game and the Race for Empire in Central Asia.* Washington, DC: Counterpoint, 2000. An outstanding retelling of the struggle to control the Eurasian heartland, with insights into the conflicts playing out in the region today.

The Qur'an. Trans. Abdulla Yusuf Ali. New York: Tahrike Tarsile Qur'an Inc, 2001. The Qu'ran was always a background presence in Central Asia, from Jalolov's pronouncements on traditional respect for water to Sabit's moralistic reflections on our predicament.

Smith, David R. "Environmental Society and Shared Water Resources in Post-Soviet Central Asia." *Post-Soviet Geography*, June 1995. Smith predicts that "water wars" will be inevitable in Central Asia if the states don't begin to cooperate.

Steyngart, Gary. *Absurdistan, A Novel.* New York: Random House, 2006.

Terzani, Tiziano. *Goodnight Mr. Lenin: A Journey through the End of the Soviet Empire.* London: Picador, 1993. A journalistic treat; Terzani takes us along the Amur River on the border of China and Siberia and to other border regions as the USSR collapses.

Robert Ferguson

Thubron, Colin. *The Lost Heart of Central Asia.* London: Heinemann, 1994. History and travels in the region in early post-Soviet days.

Whittell, Giles. *Extreme Continental: Blowing Hot and Cold through Central Asia.* London: Indigo, 1996. A spirited and amusing romp that accurately captures the area's colourful characters and extremes of beauty and despair.

Wines, Michael. "Grand Soviet Scheme for Water in Central Asia is Foundering." *New York Times* 9 Dec. 2002

Acknowledgements

This book owes its existence to the BDPA team I worked closely with in Central Asia in 2000, especially to Victor Tsoy, Irina Vovchenko, Sabit Madaliev and Jean-Charles Torrion. I am also indebted to the "opposing team" of Kadirbek Bozov and Nadir Maksumov, their five national team leaders and Rim Guiniyatullin: while it was never my intention to compete so recklessly with them – or even to compete with them at all – it was their enthusiasm for the game that really made this story.

A special recognition goes to Shakhlo Abdullayeva: whatever her transgressions, her pluck, tenacity and audacity were dazzling, and she never deserved her grim fate.

I also thank the following for their insights, critiques, encouragement and faith in the creation of this work: Eric and Marlen in Tashkent; Jill, Dave and Armorer in London; the Abacus Gang, Jon, Eithne, Sarah and Sarah; and Paul, Tom and my nephew Graeme, all in Toronto. And especially my sister, Marnie Ferguson; my agent, Denise Bukowski; and my editor, Scott Steedman.

Iguana Books

iguanabooks.com

If you enjoyed *Dancing with the Vodka Terrorists: Misadventures in the 'Stans* ...
Look for other books coming soon from Iguana Books! Subscribe to our blog for updates as they happen.

iguanabooks.com/blog/

You can also learn more about Rob Ferguson and his upcoming book *Jim Smith* on his blog.

robferguson.iguanabooks.com/blog/

If you're a writer ...
Iguana Books is always looking for great new writers, in every genre. We produce primarily ebooks but, as you can see, we do the occasional print book as well. Visit us at iguanabooks.com to see what Iguana Books has to offer both emerging and established authors.

iguanabooks.com/publishing-with-iguana/

If you're looking for another good book ...
All Iguana Books books are available on our website. We pride ourselves on making sure that every Iguana book is a great read.

iguanabooks.com/bookstore/

Visit our bookstore today and support your favourite author.

IGUANA

www.ingramcontent.com/pod-product-compliance
Lightning Source LLC
Chambersburg PA
CBHW070031100426
42740CB00013B/2653